XVII International Youth Scientific School on Actual Problems of Magnetic Resonance and its Applications 2014

Journal of Physics: Conference Series Volume 560

Kazan, Russia
22-27 June 2014

Editors:

M.S. Tagirov
A.S. Nizamutdinov
V.V. Semashko

ISBN: 978-1-63439-669-1
ISSN: 1742-6588

Printed from e-media with permission by:

Curran Associates, Inc.
57 Morehouse Lane
Red Hook, NY 12571

CURRAN ASSOCIATES INC. proceedings.com

Some format issues inherent in the e-media version may also appear in this print version.

Printed by Curran Associates, Inc. (2015)

Institute of Physics
Dirac House, Temple Back
Bristol BS1 6BE UK

Phone: 44 1 17 929 7481
Fax: 44 1 17 920 0979

techtracking@iop.org

Additional copies of this publication are available from:

Curran Associates, Inc.
57 Morehouse Lane
Red Hook, NY 12571 USA
Phone: 845-758-0400
Fax: 845-758-2634
Email: curran@proceedings.com
Web: www.proceedings.com

XVII International Youth Scientific School on Actual Problems of Magnetic Resonance and its Applications 2014

Journal of Physics: Conference Series Volume 560

Kazan, Russia
22-27 June 2014

ISBN: 978-1-63439-669-1
ISSN: 1742-6588

IOP Institute of Physics

Table of contents

Volume 560

XVII International Youth Scientific School on Actual Problems of Magnetic Resonance and its Applications
22–27 June 2014, Kazan, Russia

Accepted papers received: 5 November 2014
Published online: 24 November 2014

Preface

Papers

Journal of Physics: Conference Series **560** (2014) 011001 doi:10.1088/1742-6596/560/1/011001

Preface

Editors: M.S.Tagirov, V.V.Semashko, A.S.Nizamutdinov

Kazan is the motherland of Electronic Paramagnetic Resonance (EPR) which was discovered in Kazan State University in 1944 by prof. E.K.Zavojskii. Since the Young Scientist School of Magnetic Resonance run by professor G. V. Skrotskii from MIPT stopped its work, Kazan took up the activity under the initiative of academician A. S. Borovik-Romanov. Nowadays this school is rejuvenated and the International Youth Scientific School studying “Actual problems of the magnetic resonance and its application” is developing. Traditionally the main subjects of the School meetings are: Magnetic Resonance in Solids, Chemistry, Geology, Biology and Medicine. The unchallenged organizers of that school are Kazan Federal University and Kazan E. K. Zavoisky Physical -Technical Institute. The rector of the School is professor Murat Tagirov, vice-rector – professor Valentine Zhikharev.

Since 1997 more than 100 famous scientists from Germany, France, Switzerland, USA, Japan, Russia, Ukraine, Moldavia, Georgia provided plenary lecture presentations. Almost 700 young scientists have had an opportunity to participate in discussions of the latest scientific developments, to make their oral reports and to improve their knowledge and skills. To enhance competition among the young scientists, reports take place every year and the Program Committee members name the best reports, the authors of which are invited to prepare full-scale scientific papers.

Since 2013 the International Youth Scientific School “Actual problems of the magnetic resonance and its application”, following the tendency for comprehensive studies of matter properties and its interaction with electromagnetic fields, expanded “the field of interest” and opened the new section: Coherent Optics and Optical Spectroscopy. Many young people have submitted interesting reports on photonics, quantum electronics, laser physics, quantum optics, traditional optical and laser spectroscopy, non-linear optics, material science and nanotechnology. Here we are publishing the full papers prepared from the most interesting reports selected by the Program Committee of the School.

The XVII International Youth Scientific School has been supported by a subsidy from the Russian Government (agreement #02.A03.21.0002) in frame of Program of Competitive Growth of Kazan Federal University among World’s Leading Academic Centers. It is a pleasure to thank the sponsors (BRUKER Ltd, Moscow, the Russian Academy of Science, the Dynasty foundation of Dmitrii Zimin, Russia, Russian Foundation for Basic Research and R. M. Minnikhanov, president of Republic of Tatarstan) and all the participants and contributors who made the International School meeting possible and interesting.

Peer review statement

All papers published in this volume of *Journal of Physics: Conference Series* have been peer reviewed through processes administered by the proceedings Editors. Reviews were conducted by expert referees to the professional and scientific standards expected of a proceedings journal published by IOP Publishing.

Journal of Physics: Conference Series **560** (2014) 012001 doi:10.1088/1742-6596/560/1/012001

Growth of solid solutions with colquiriite structure $LiCa_{0,2}Sr_{0,8}AlF_6$: Ce^{3+}

A A Shavelev, A S Nizamutdinov, V V Semashko, M A Marisov

Kazan Federal University, 420008, Kremlevskaja str., 18, Kazan, Russian Federation

E-mail: alexey.shavelev@gmail.com

Abstract. Aim of this work were experiments on growing new materials based on fluoride crystals with the colquiriite structure $LiSr_{0,8}Ca_{0,2}F_6$, as well as the study of their phase composition. It is shown that for a series of crystals $LiSr_{0,8}Ca_{0,2}F_6$ distribution of reflections observed corresponds to the colquiriite structure, and the dependence of the lattice constant in the transition from $LiCaAlF_6$ crystal to $LiSrAlF_6$ crystal is linear. Also it found that absorption coefficient in mixed samples is much larger than in not mixed.

1. Introduction

Rare-earth doped $LiCaAlF_6$ (LiCAF) and $LiSrAlF_6$ (LiSAF) crystals has been investigated for various optical and scintillation applications. Especially, Ce doped LiCAF and LiSAF crystals have been gathering attentions as a novel neutron scintillator material with a good sensitivity to thermal neutron and fast decay time [1,2,3]. The advantages of these compounds over oxide crystals - the melting point is significantly lower (810°C) and the band gap is much wider (11 eV). Significant inhomogeneous broadening of the vibrational laser transitions of these ions causes a broad amplification band which gives ability to wavelength tuning and ultrashort pulses generation [4, 5].

Ce^{3+} activators in Ce:LiCAF are supposed to substitute sites occupied by Ca^{2+}, which has an ionic radius similar to that of Ce^{3+} (1Å and 1,03 Å respectively) [1]. However, it is difficult to increase the Ce^{3+} concentration in Ce:LiCAF crystals because of the difference between the valences of Ce^{3+} and Ca^{2+} ions, which can give rise to defects nearby to preserve the electroneutrality of the crystal. Consequently, the segregation coefficient of Ce^{3+} ions in Ce:LiCAF crystals is quite small [4]. It is known, that varying the chemical structure of the compound, namely a set of crystal lattice cations, one can improve the optical quality by increasing the isomorphic capacity of solid solution [6].

The aim of this work are experiments on growing new materials based on mixed fluoride crystals with the colquiriite structure $LiSr_{0,8}Ca_{0,2}F_6$ (LiSCAF) as well as the study of their phase composition.

2. Conditions, characteristics and growing complexity of the materials.

A special role in the successful use of fluorides plays chemical purity of crystalline samples: removing traces of oxygen and hydroxyl groups in the crystals concomitant reduction of impurities - the rare earth ions and ions of the iron group. Furthermore, the final performance of the solid-state laser crystal material is affected strongly by its optical quality and the number of defects of crystal structure which in turn is influenced by especially the crystal growth and subsequent machining. In particular, one of the main problems arising from the operation of materials under intense UV radiation is the degradation of their optical properties.

As a method of crystal growth the Bridgman-Stockbarger was selected, where crystals are grown in closed crucibles. This maintains the chemical composition of the melt and allows to produce crystal samples with various activator ions even for the case of extremely low segregation coefficients of demanded impurities, which takes place for investigated Ce^{3+}:LiCAF crystals.

The melting point of the crystals comprise 810 ^{0}C [7]. These systems are multicomponent, and, as indicated above, are characterized by an extremely low segregation coefficient of Ce^{3+} ions which are active ions of the gain medium. Therefore, it is critical to arrange a sufficient temperature gradient in the crystallization zone during growth of these crystals [8]. The temperature gradient is determined by the geometry of the following elements: conductor current, heater, heat shields and crucible.

Complexity of growing is in the fact that aluminum is volatile, therefore: it is not allowed to overheat the melt, since the components include the aluminum fluoride, which is a compound with a high saturated vapor pressure and does not exist in a liquid state, so overheating of the melt will lead to depletion of this component. Also, when the temperature is set taking the value with an excess (about 20^{0}C) to ensure melting of the substance.

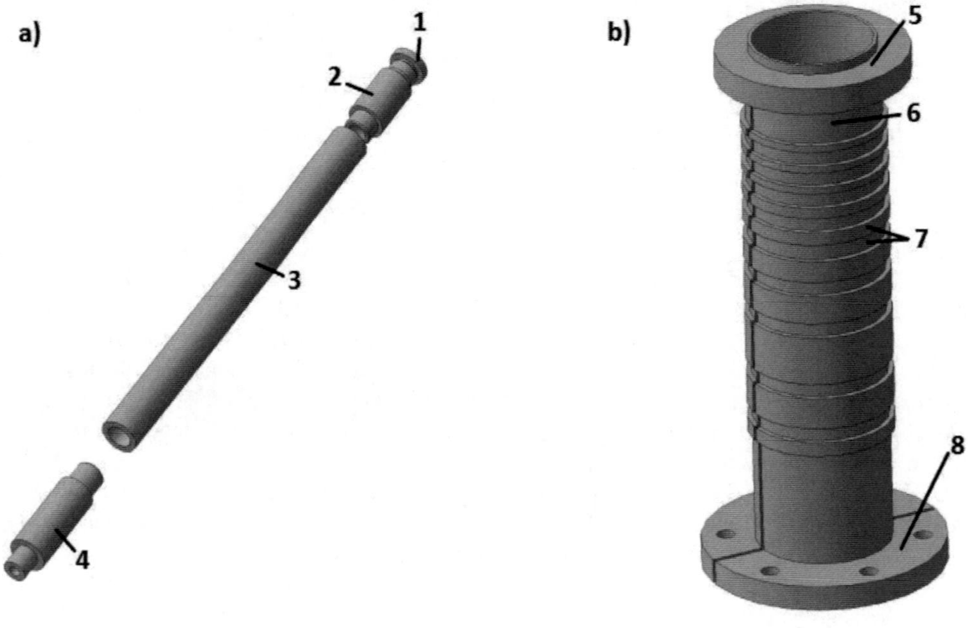

Figure 1. Crucible (a) and the heater (b) of the growth chamber. Crucible consists of: 1) lid, 2) container for teflon, 3) container for charge, 4) the pin in which put the seeding crystal. b) body of resistive heater construction: 5) place for shield, 6) heater, 7) variable thickness to form a gradient field, 8) mounting place for coupling to the current conductor.

Crucible also participates in the formation of a temperature gradient. The wall thickness of the crucible should be minimal (in order 2-3mm) to distort the temperature field as little as possible.

Series of crystals Ce^{3+}:LiCAF: and Ce^{3+}:LiSCAF with 1 at. % content of CeF_3 in the initial charge were grown with Bridgman-Stockbarger method. A fine powder samples were prepared for investigation by X-Ray diffractometry from the synthesized material. Also samples for optical spectroscopy were prepared from the synthesized boules. Polished plane-parallel windows were made along the boule of Ce:LiCAF crystal. For the Ce:LiSCAF crystal plane-parallel plates with polished windows were prepared. The optical c-axis was kept in the plane of polished windows. Length of the optical path in crystals Ce:LiCAF and Ce:LiSCAF was 8 mm and 2 mm respectively.

Journal of Physics: Conference Series **560** (2014) 012001 doi:10.1088/1742-6596/560/1/012001

3. Diffractometry results

Phase composition of crystallized material was investigated by means of X-ray difractometry technique. The correspondent reflections distributions are shown at Fig. 2.

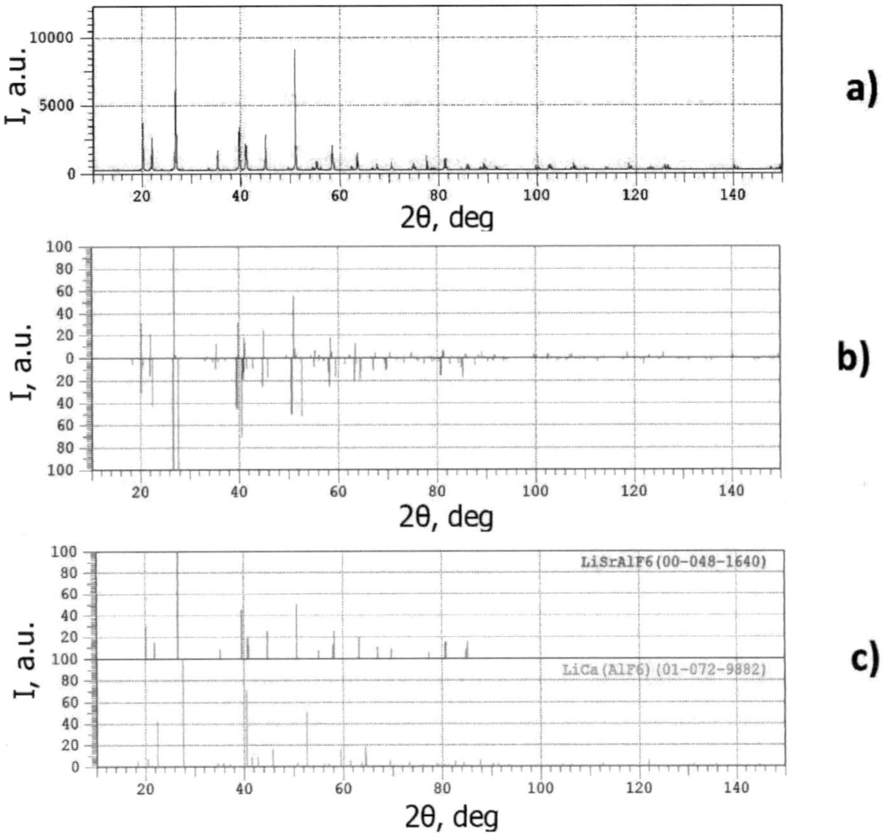

Figure 2. X-ray diffraction patterns of the samples: LiSCAF:Ce^{3+} experimental data and LiCAF and LiSAF from the works [9, 10]: a) experimental data, b) comparison data, c) table data.

The diffraction pattern obtained for Ce^{3+}:LiSCAF corresponds to colquiriite structure. From comparison to LiCAF and LiSAF data [9, 10] it is seen that Ce^{3+}:LiSCAF reflections are lying between LICAF and LISAF data as it was expected due to shift of lattice constant.

We have also obtained the diffraction lattice constant dependence on the concentration ratio of mixed and unmixed samples (Figure 3).

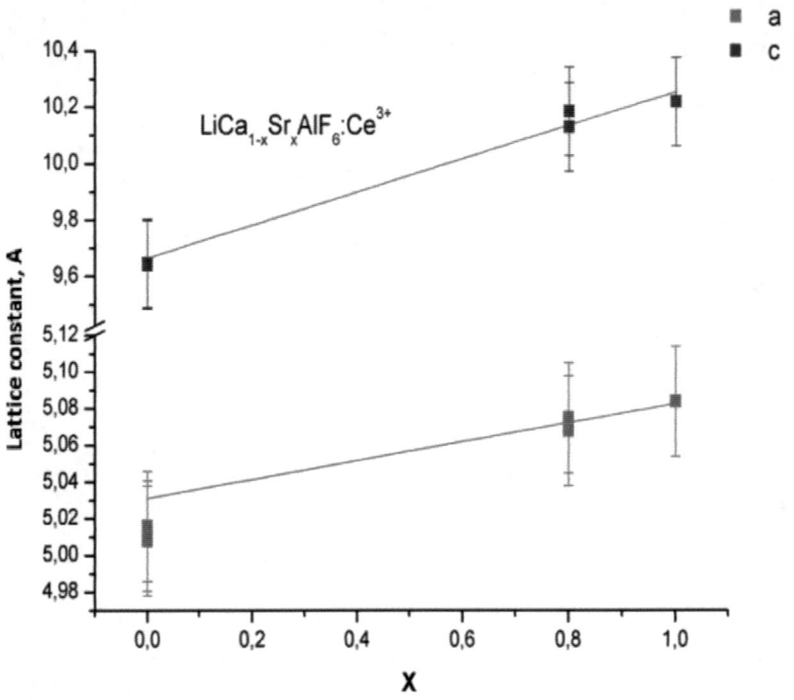

Figure 2. Dependence of the lattice constants **a** and **c** of crystallized Ce^{3+}:LiSCAF material on the SrF_2 content.

We see that lattice constants for Ce^{3+}:LiSCAF lie on the straight lines between LICAF and LISAF and this means that our crystal is monocrystalline and has the only phase according to Vegard's rule.

4. Absorption spectra results

At Fig. 4 the absorption bands corresponding to lowest 4f-5d transition of Ce^{3+} ion in are presented. In the absorption spectrum we see a small blue shift of the band. But we remind you that this is a superposition of a number of spectrum bands corresponding to different nonequivalent impurity centers of Ce^{3+} [11]. An important result is that the absorption coefficient of the mixed sample about 5 times larger than the absorption coefficient of not mixed sample (Figure 4).

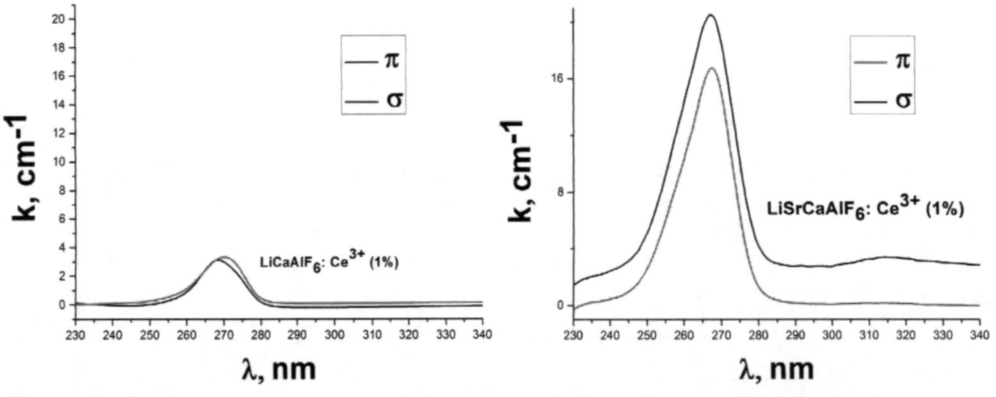

Figure 3. Absorption bands of Ce^{3+}:LiCAF and Ce^{3+}:LiSCAF in different polarizations.

Journal of Physics: Conference Series **560** (2014) 012001 doi:10.1088/1742-6596/560/1/012001

5. Conclusion

Solid solutions with chemical formula $LiSr_{0,8}Ca_{0,2}F_6$ doped with Ce^{3+} ions were grown. It is shown that for a series of crystals $LiSr_{0,8}Ca_{0,2}F_6$ distribution of X-ray diffraction reflections observed corresponds to the colquiriite structure, and the dependence of the lattice constant in the transition from $LiCaAlF_6$ crystal to $LiSrAlF_6$ crystal is linear. These facts indicate that the synthesized crystals are single-phase crystals. Also it was found that absorption coefficient in mixed samples is much larger than in not mixed and this speaks for higher distribution coefficient of Ce^{3+} ions in mixed crystals.

Acknowledgements

This work was funded by the subsidy allocated to Kazan Federal University for the project part of the state assignment in the sphere of scientific activities.

References

[1] Yokota Y, Fujimoto Y et al 2014 *IEEE Transactions on Nuclear Science* **614** 19-423

[2] Wellmann B, Spence D, and Coutts D Tunable continuous-wave deep-ultraviolet laser based on Ce: LiCAF 2014 *OPTICS LETTERS* **39** 1306-1309

[3] Thanh L, Schowalter S et al Low-threshold ultraviolet solid-state laser based on a Ce^3:$LiCaAlF_6$ crystal resonator 2012 *OPTICS LETTERS* **37** 4961-4963

[4] Klimm D, Uecker R, Reiche P 2005 *Cryst. Res. Technol* **40** 352-358

[5] Dubinskii M.A., Semashko V.V., Naumov A.K., Abdulsabirov R. Y., Korableva S. L. 1993 *J. Modern Opt* **40** 1-5

[6] Sarukura N., Liu Z., Ohtake H., Segawa Y., Dubinskii M.A., Semashko V.V., Naumov A.K., Korableva S.L., Abdulsabirov R. Yu. 1997 *Opt.* **22** 994-996

[7] Shimamura, K. Growth of Ce-doped $LiCaAlF_6$ and $LiSrAlF_6$ single crystals by the Czochralski technique under CF4 atmosphere [Text] 2000 *J. Crystal Growth* **211** 302

[8] Nizamutdinov A.S., Semashko V.V., Naumov, A.K., Efimov V.N., Korableva S.L., Marisov M.A. 2010 *JETP LETTERS* **91** 23-25

[9] Kuze, S., Du Boulay et al Structures of $LiCaAlF_6$ and LiSrAlF6 at 120 and 300 K by synchrotron X-ray single-crystal diffraction 2004 *Journal of Solid State Chemistry* **177** 3505-3513

[10] Rupp, B., Kway et al Chromium Site Specific Chemistry and Crystal Structure of $LiCaCrF_6$ and Related Chromium Doped Laser Materials 1993 *Journal of Solid State Chemistry* **107** 471-479

[11] Semashko V V, Dubinskii M A, Abdulsabirov R Yu, Naumov A K, Korableva S L,Shcherbakova N K, and Klimovitskii A E Investigation of Multisite Activation in LiCaAlF6: Ce^{3+} Crystals Using Stimulated Quenching of Luminescence 1995 *Laser Physics* **5** 69.

Ultrashort pulsed UV lasers based on the Ce^{3+}:$LiCaAlF_6$ and $LiLuYF_4$:Ce^{3+} crystals

O R Akhtyamov, V V Semashko, A S Nizamutdinov, M A Marisov

Kazan Federal University, 420008, Kremlevskaja str., 18, Kazan, Russian Federation

E-mail: akhtyamovo@mail.ru

Abstract. Here we report on UV laser oscillation on colquiriite structure fluoride mixture crystals of Ce^{3+}:$LiCaAlF_6$ (Ce:LiCAF) and Ce^{3+}:$LiLuYF_4$ (Ce:LiLuYF) active media. We report on 800 ps pulse train lasing form Ce:LiCAF in composite cavity and single pulse lasing of 400 ps duration from Ce:LiLuYF in convenient Fabri-Perrot cavity and discuss conditions of pulse shortening in UV active medium by photodynamic processes.

An important problem of quantum electronics is the generation of laser radiation in the form of ultrashort pulse in UV spectral range. The family of cerium-doped laser crystals [1] offer an attractive all-solid-state route providing valuable tunable laser oscillation in the UV directly with simpliness and attractive properties of solid-state laser (beam divergence, energy distribution) which is problematic for nonlinear converters. Here we report on UV laser oscillation on colquiriite structure fluoride mixture crystals of Ce^{3+}:$LiCaAlF_6$ (Ce:LiCAF) and Ce^{3+}:$LiLuYF_4$ (Ce:LiLuYF) active media which can be pumped at around 266 nm, and so can utilize frequency-quadrupled Nd-based pump lasers [2].

For the first time lasing on the Ce:LiCAF crystal was reported by Dubinskii [3]. Broad gain band of Ce:LiCAF crystal from 280 to 315 nm allows amplification and generation of femtosecond pulses. Using the 4-pass amplifier based on the Ce:LiCAF crystal Zhenlin [4] report about amplification of femtosecond pulse of 115 fs duration to 370 times at a wavelength of 290 nm.

Mathematical modeling conducted for the first time in [5] showed significant reduce of the generated pulses duration, with proper selection of the parameters of the laser cavity and the pump level. The conditions of reduced pulse duration are high exceeding of the population inversion in the active medium over the lasing threshold and large ratio of the pump pulse duration to the lifetime of photons in the cavity [5]. However, the color centers (CC) formation under pump radiation in the active medium was not considered in this paper while CC have absorption bands at the lasing wavelength and affect at its evolution.

Effect of CC on the characteristics of the Ce:LiCAF lasers were discussed earlier [6] and it has been shown that with increasing pump energy total intracavity losses decrease due to CC. This was explained by the color centers formation under the pump radiation and their subsequent destruction by the laser radiation. Color centers absorption is getting lower with higher amount of laser radiation remaining in the cavity [6].

In the present work modeling of the laser oscillation was carried out considering CC formation and the mechanism of laser pulse shortening due to CC was investigated.

The model contains conventional four-level scheme, the conduction band of the crystal and the CC level (fig. 1).

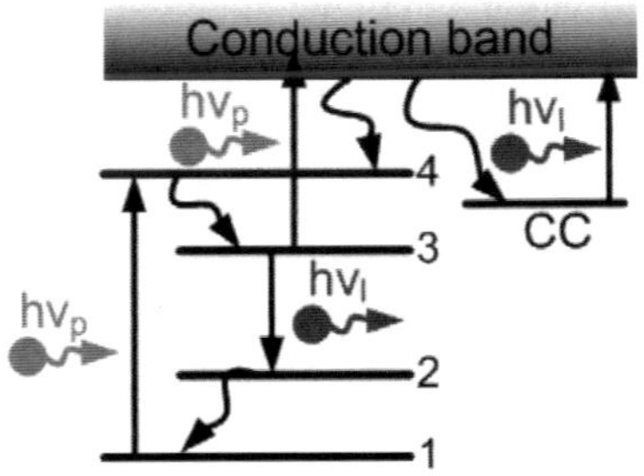

Figure 1. The level scheme of the UV active medium corresponding to four-level lasing scheme and formation of color centers Crucible also participates in the formation of a temperature gradient. The wall thickness of the crucible should be minimal (in order 2-3mm) to distort the temperature field as little as possible.

In accordance with this scheme system of differential equations were written (1-8), which are solved by computer software Matlab. Lasing simulation results are shown in Figure 2.

$$\frac{dn_1(t)}{dt} = N_{pump}(t)\sigma_{14}\left(n_4(t) - n_1(t)\right) + \frac{n_2(t)}{t_{21}} \tag{1}$$

$$\frac{dn_2(t)}{dt} = q(t)\sigma_{32}\left(n_3(t) - n_2(t)\right) - \frac{n_2(t)}{t_{21}} + \frac{n_3(t)}{t_{32}} \tag{2}$$

$$\frac{dn_3(t)}{dt} = \frac{n_4(t)}{t_{43}} - q(t)\sigma_{32}\left(n_3(t) - n_2(t)\right) - \frac{n_3(t)}{t_{32}} - \sigma_{ESA}N_{pump}\left(n_3(t) - n_{CB}(t)\right) \tag{3}$$

$$\frac{dn_4(t)}{dt} = -N_{pump}(t)\sigma_{14}\left(n_4(t) - n_1(t)\right) - \frac{n_4(t)}{t_{43}} + \frac{n_{CB}(t)}{t_{CB}} \tag{4}$$

$$\frac{dn_{CB}(t)}{dt} = N_{pump}(t)\sigma_{ESA}\left(n_3(t) - n_{CB}(t)\right) - \frac{n_{CB}(t)}{t_{CB}} - \frac{n_{CB}(t)}{t_{CBCC}} + \frac{n_{CC}(t)}{t_{CCCB}} + \sigma_{CC}n_{CC}(t)\left(q(t) + \frac{n_3(t)}{t_{32}}\right) \tag{5}$$

$$\frac{dn_{CC}(t)}{dt} = \frac{n_{CB}(t)}{t_{CBCC}} - \sigma_{CC}n_{CC}(t)\left(q(t) + \frac{n_3(t)}{t_{32}}\right) - \frac{n_{CC}(t)}{t_{CCCB}} \tag{6}$$

$$\frac{dq(t)}{dt} = q(t)\left(\sigma_{32}\left(n_3(t) - n_2(t)\right) - \sigma_{CC}n_{CC}(t)\right)\frac{c}{L} - \frac{q(t)}{t_c} \tag{7}$$

$$N_{Ce} = n_1 + n_2 + n_3 + n_4 + n_{CC} + n_{CB} \tag{8}$$

Where n_1 – n_4, n_{CB}, n_{CC} – corresponding population on Ce^{3+} energy levels, CC and conduction band; N_{pump} and q – pump and lasing photon flux density; σ_{14} and σ_{ESA} – absorption cross section and excited state absorption cross section on pump wavelength; σ_{32} and σ_{CC} – gain cross section and CC absorption cross section on lasing wavelength; $\frac{1}{t_{21}}, \frac{1}{t_{32}}, \frac{1}{t_{43}}, \frac{1}{t_{CB}}, \frac{1}{t_{CBCC}}, \frac{1}{t_{CCCB}}, \frac{1}{t_{CC}}$ – probabilities of the corresponding transitions; c – speed of light; L – cavity length; N_{Ce} – concentration of Ce^{3+} ions.

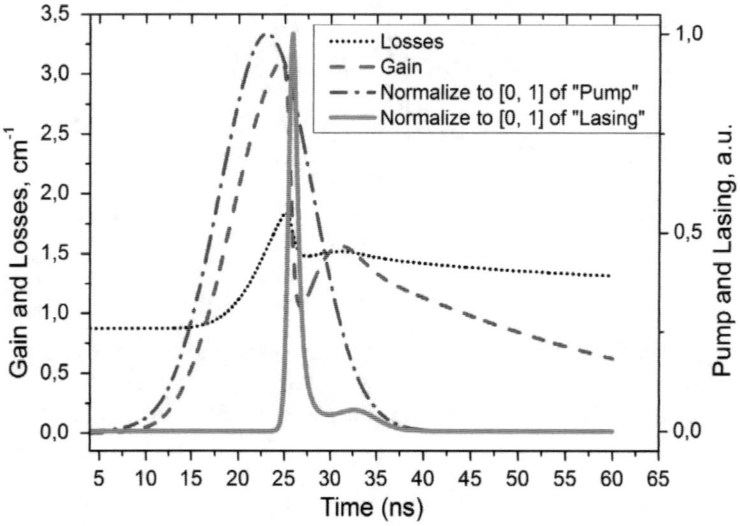

Figure 2. The simulation results considering photodynamic processes

In the simulation procedure the parameters of the active medium and short cavity were considered corresponding to low Q-factor. As a result, we have obtained the case of single ultrashort pulse generation with duration of 400 ps. According to the CC populations graph (black curve named "losses") it is clear that CC in the bulk this medium work as saturable absorber (SA). Namely pump radiation induces the CC formation, which leads to higher losses thus forms the SA. When laser radiation appears it starts bleaching the CC and this process corresponds to saturation of SA. This leads to sharp increase in Q-factor of the cavity and determines conditions for the ultrashort pulses generation.

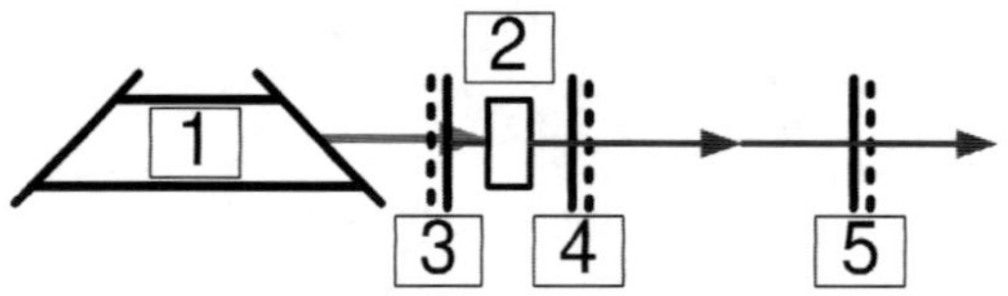

Figure 3. Laser experiment scheme: 1 – pump source (Nd :YAG 266 nm / Raman-shifter 300 nm); 2 – active medium (Ce:LiCAF / Ce:LiLuYF); 3, 4 – flat mirrors (R=25% on lasing wavelength), cavity length 1,5 cm; 5 – extra mirror of the cavity.

In order to test this mechanism of laser pulse shortening the experiments were carried out in a short low-Q cavity with Ce:LiCAF and Ce:LiLuYF crystals as active elements. The experimental scheme is shown in Figure 3.

We have carried out experiments on generation of subnanosecond pulses on Ce:LiCAF crystal with 266 nm 10 ns pumping form Nd:YAG laser. In these experiments we have not achieved stable generation of single ultrashort pulse at any pump energy but using composite cavity we have achieved stable “pulse-train” generation. Experimental results are presented in fig.4a.

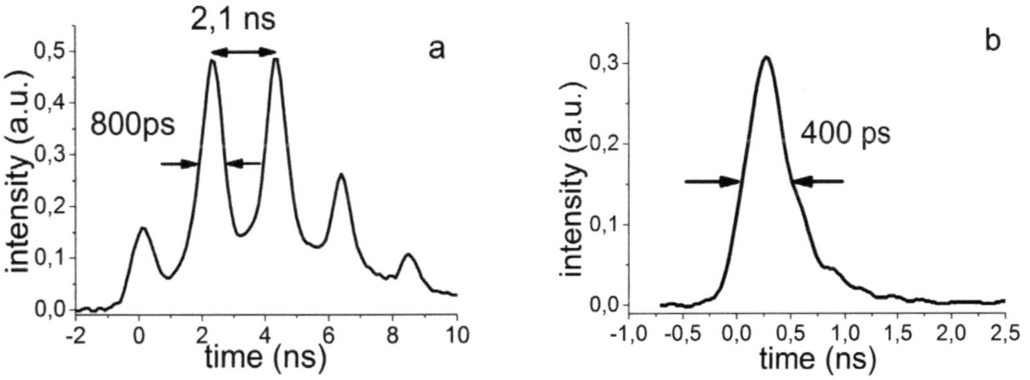

Figure 4. Temporal distribution Ce:LiCAF laser pulses (with external mirror) (a); temporal distribution Ce:LiLuYF laser pulse (b)

The composite laser cavity consists of a short length low Q-factor pulse-seeding laser cavity and a feedback laser cavity. The pulse-seeding laser cavity provides only spiking mode of laser oscillation. Regenerative amplification was organized by aligning the axes of seeding cavity and long feedback cavity [1,2], and as a result we see a stable ultrashort "pulse-train" with pulse duration at about 800 ps and a period defined by the design of the cavity. It is important that length of the external cavity should be tuned so that the second pulse of "pulse-train" occur before generation of the second stochastic pike. In this case the lasing pulse is circulating inside the external laser cavity harvests the inversion and regenerative amplification prevents stochastic laser oscillation.

In a laser based on Ce:LiCAF crystal the mechanism of laser pulse shortening due to CC is not working, possibly due to the low level of losses associated with the CC formation. However, losses in the Ce:LiLuYF crystal associated with CC may exceed the same losses in the Ce:LiCAF crystal 5 times [6,7].

In experiments using LiLuYF crystal in low-Q short cavity we have obtained a single ultrashort pulse generation of 400 ps duration with pump radiation at the wavelength 300 nm obtained from Raman-shifter pumped by Nd:YAG laser. Therefore, we have proved the possibility of use of photodynamic processes for intracavity losses modulation. Internal UV pump induced losses in active medium were utilized to shorten the laser pulse.

Acknowledgements
This work was funded by the subsidy allocated to Kazan Federal University for the project part of the state assignment in the sphere of scientific activities.

References
[1] Sarukura N, Liu Z, Segawa Y, Semashko V V et al 1995 *J. Appl. Phys. Lett.* **67** 602
[2] Sarukura N, Liu Z, Izumida S, Dubinskii M A, Abdulsabirov R Y, Korableva SL 1998 *J. Appl. Opt.* **37** 6446
[3] Dubinskii M A, Abdulsabirov R Y, Korableva S L, Naumov A K, Semashko V V 1993 *Proc. OSA on Advanced Solid-State Lasers* **15** 195-198
[4] Zhenlin L, Toshimasa K, Yuji S, Nobuhiko S, Kiyoshi S, Tsuguo F, Masahiro H and Hideo H 2001 *Opt. Lett.* **26** 301-303
[5] Roess D 1966 *J.Appl.Phys.* **37** 2004-2006
[6] Akhtyamov O R, Nizamutdinov A S, Semashko V V, Naumov A K, Korableva S L 2011 *Proc. SPIE.* **7994** 79940I
[7] Semashko V V, Dubinskii M A, Abdulsabirov R Y, Korableva S L, Naumov A K, Nizamutdinov A S, Zhuchkov M S 2002 *Proc. SPIE.* **4766**

Laser-like effects and upconversion fluorescence temporal dynamic in Tm^{3+}, Yb^{3+} doped YF_3 single crystals

B N Kazakov, V V Semashko, A V Lovchev, A K Naumov

Kazan Federal University, 420008, Kremlevskaja str., 18, Kazan, Russian Federation

E-mail: bor300@yandex.ru

Abstract. A new way of managing fluorescence features in doped bulk and nanosized materials by variation of excitation temporal characteristics is demonstrated. Random laser-like effect on $^3H_4$-$^3F_4$ transitions of Tm^{3+} ions in Yb,Tm:YF_3 single crystal was revealed under rectangular pulse train excitation in 930-980 spectral domain.

1. Introduction

Modification of optical excitation conditions allows the effective managing of upconversion (UC) processes and gives an opportunity to populate or depopulate high-lying states of doped bulk and nanosized materials. This is fruitful for realization of multicolor lasers and high-effective luminophors for IT [1], biomedicine [2] and photovoltaic applications [3]. The most typical physical mechanisms determining distribution of population of active ions energetic levels under UC excitation are the processes combination: a) cross-relaxation (CR), b) excited-state absorption (ESA) of excitation and/or fluorescence radiations, c) energy transfer and energy migration between dopants, d) cooperative processes [4]. As it is generally considered, materials' UC-fluorescence features are operated by excitation radiation wavelength and its intensity or by changing the chemical composition of samples (e.g. by dopant types and/or dopant concentrations) or by variation of sample size characteristics or ambient conditions (e.g. by sample's temperature, pressure, additional electro-magnetic fields etc.). However there is still a forgotten additional "degree of freedom", which impacts on population evolution of excited state of active ions – it is exposure time (pulses length, their repetition rate or different types excitation pulse trains). For the first time to our knowledge here we demonstrate opportunity to manage spectral-fluorescence and kinetics properties of YF_3 single crystals doped by Yb^{3+} (20 at.%) и Tm^{3+} (1 at.%) ions (Yb,Tm:YF_3) by means of variation of optical excitation exposure time in $^2F_{7/2}$-$^2F_{5/2}$ transitions domain of Yb^{3+} ions.

2. Experimental peculiarities and results

Multiple rare-earth ions doped YF_3 crystal host is well known as high effective phosphor due to low probability of multiphonon non-radiative processes and more close packing of the active ions in comparison with double- and triple fluoride hosts [5,6]. Orthorhombic YF_3 single crystals doped by Yb^{3+} and Tm^{3+} were grown by Bridgman-Stockbarger technique in Kazan Federal University. Yb^{3+} and Tm^{3+} ions contents in the crystal were 20 and 1 at.%, accordingly. Because of crystallization peculiarity, samples look as the needles extended along C axis with diameter up to 2 mm and up to 3-5 mm long.

The Yb,Tm:YF_3 samples fluorescence was excited by radiation of temperature tuned laser emitting diodes (LED) operated around 930 and 980 nm. This wavelenegth is correspondent to $^2F_{7/2}\rightarrow{}^2F_{5/2}$ transition of Yb^{3+} ions. Maximal LED output power was about 4 W. LED provided a sequence of rectangular laser pulses with pulse repetition rate 66.7 Hz and varied pulse duration τ_{ex} = 0.08÷14.5 ms. LED radiation was collimated by spherical lens to slightly distorted circle spot about 0.25 mm in diameter. Fluorescence spectra were registered by CCD StellarNet spectrometer-radiometer or by scanning spectrometer based on prismatic ZMR-3 monochromator. Both spectrometers provided spectral resolution better than 0.5 nm [7]. Fluorescence decays were registered by means of photomultipliers FEU-39, FEU-79 and FEU-62 with temporal resolution better than 1-5 μs. Input window of photomultiplier was attached to output slit of ZMR-3 monochromator and electrical output of the photomultiplier was connected to 12-bit high-speed 4 channel A/D convertor L-Card E20-10 with 30 MHz bandwidth. The principal feature of decay measurements was to avoid data averaging and study every response of Yb,Tm:YF_3 system for IR excitation act allowing detection of new fluorescence dynamic peculiarities. All experiments were carried out at 300 K.

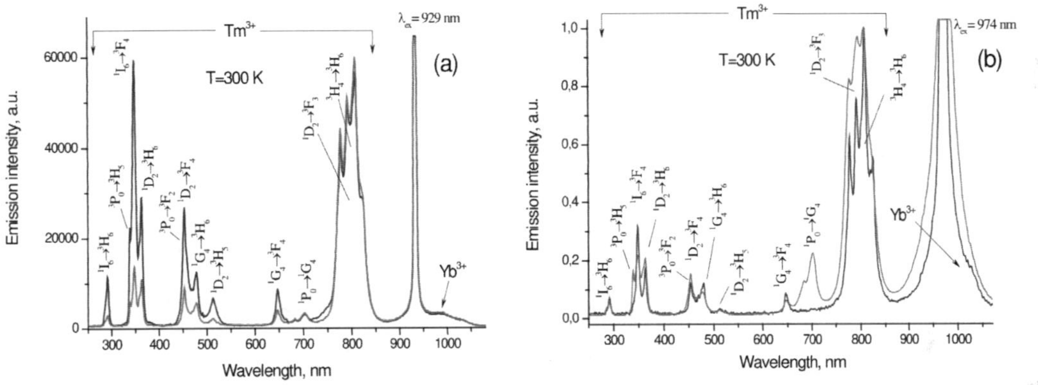

Figure 1. Typical upconversion fluorescence spectra of Tm^{3+} ions in double-doped Yb,Tm:YF_3 crystal under excitation of Yb^{3+} ions at 929 and 974 nm by CW (a) and by rectangular pulses train (b). Spectra marked by blue and red lines were registered under CW excitation by 929 nm radiation with about 90 and 8000 W/cm^2 power density (a) and under sequences of rectangular 1 ms and 14 ms LED pulses (pulse repetition rate 66.7 Hz, λ_{ex}=974 nm, power density ~ 200 W/cm^2) (b), correspondingly.

Intensive UC-fluorescence originated from ^3P_j, $^1I_6$, $^1D_2$, $^1G_4$, $^3F_3$, $^3H_4$ states of Tm^{3+} ions were detected under excitation of Yb^{3+} ions at 920-980 nm spectral domain. The processes which lead to such UC-fluorescence features are known well [4-7] and have to be considered as dynamic. But there are neither comprehensive experimental nor theoretical studies to our knowledge aimed at demonstration of the opportunity to utilize these temporal dynamics to population managing of activator ions higher-lying manifolds. Indeed it is pronounced well in UC-fluorescence spectra of double-doped Yb,Tm:YF_3 single crystals under various excitation temporal characteristics. For instance, typical fluorescence spectra of Yb^{3+} and Tm^{3+} ions in YF_3 crystal under continues wave (CW) (a) and under sequence of rectangular pulses (b) excitation of Yb^{3+} ions at 929 and 972 nm are shown on the fig.1. The fig.2 demonstrates dependences intensity of main upconversion fluorescence lines versus LED pumping current (it is identical to excitation power) (a) and versus laser excitation pulse length (b). Correspondent experimental conditions are shown in figures captions.

As it can be seen from the fig.1 and 2, variation of temporal characteristics of excitation (pulses length and/or pulse repetition rate) as well as a optical excitation intensity change lead to effective UC population of higher lying manifolds of Tm^{3+} ions.

At the same time population distribution features of these states under rectangular pulse train excitation are essentially different in comparison to CW or periodical short pulse excitation, because

UC processes include numerous intermediate energetic states of activator and/or sensitizer ions with the proper lifetimes. In particular, it is easy to see from the figures, that the conversion from pulse train to CW excitation demonstrates the rise of effective population of $^1P_0$ manifold of Tm^{3+} ions in Yb,Tm:YF_3 single crystal and appearance of intensive fluorescence bands at 339 ($^1P_0$-$^3H_5$), ~ 683 and 702 nm ($^1P_0$-$^1G_4$). The same effect takes place in case of other ^1P_j and $^1I_6$ manifolds. Therefore, sensitivity of population to excitation prehistory in similar fluorescent materials opens additional opportunity of managing the population distribution at energy states of dopants and it appears to be extremely important from the point of view of photonics urgent problems.

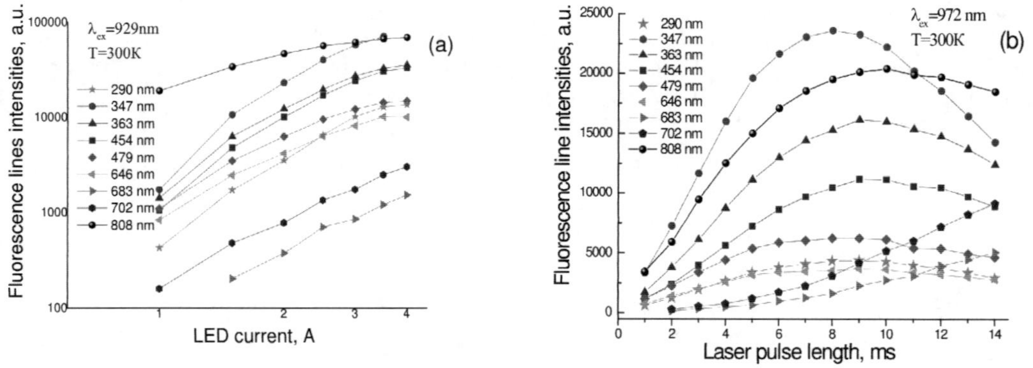

Figure 2. Dependences of intensities of upconversion fluorescence lines in Yb,Tm:YF_3 crystal versus LED pumping current (CW excitation at 929 nm) (a) and versus laser pulses length (rectangular pulse train excitation at 972 nm with the power density ~ 200 W/cm^2) (b)

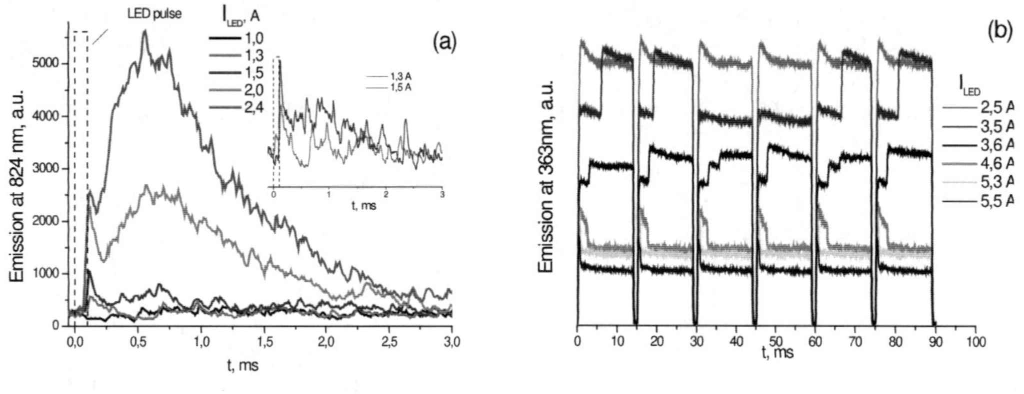

Figure 3. Random spike-likelasing distorted fluorescence decays at 824 nm ($^3H_4$-$^3H_6$ transition of Tm^{3+} ions) (a) and the example of jump-like nature of fluorescence kinetic associated with $^1D_2$-$^1H_6$ transitions of Tm^{3+} ions in Yb,Tm:YF_3 single crystals (b). Excitation pulse lengths duration was 80 μs (a) and 14 ms (b), correspondingly. Pulse repetition rate is 66.7 Hz

In addition our experiments have elucidated laser-like effects which appeared in fluorescence kinetics under rectangular pulse train excitation. For instance, random spike-like lasing distorted decay of 824 nm fluorescence originating from $^3H_4$ manifold of Tm^{3+} ions is shown on fig.3,a. Taking into account that well-known $^3H_4$-$^3F_4$ laser transition of Tm^{3+} ions is characterized by high stimulated emission cross-section [4], we propose that random lasing is appeared only at the said transition at 1.5 μm spectral domain. Despite the fact that we did not study here spectral-kinetic properties of Yb,Tm:YF_3 samples at wavelengths longer 1100 nm, the obvious evidences of such random lasing the spike-like features have well pronounced in fluorescence kinetics at 824 nm, and these features appeared above certain threshold (about 1.1-1.2 A of LED current). Moreover the stimulated emission driven character of fluorescence evaluation curve is proved by results of simulation of Tm^{3+} ions transitions with excited-state absorption and energy transfer originated from Yb^{3+} ions [8]. Jump-like modulation characteristic for $^1H_4$ manifold population is also displayed in fluorescence features at other wavelengths correspondent to transitions from higher lying manifolds of Tm^{3+} ions which could be clearly seen from curve for I_{LED}=3.6 A excitation in fig. 3,b. This is due to cross-relaxation and ESA processes originating from $^3H_4$ term and populating the higher lying states [4-6]. Another evidence of lasing feature is dropped fluorescence at $^1D_2$-$^1H_6$ transitions (363 nm) when excited with excitation power I_{LED} > 2.5 A (fig.3,b) correspondent to the lasing threshold described above. The similar behavior has fluorescence kinetics associated with transitions from ^3P_j, $^1I_6$, $^1G_4$ manifolds of Tm^{3+} ions in Yb,Tm:YF_3 posessed the similar behavior. Detailed studies of these effects are in progress.

3. Conclusion

Here we have demonstrated a new way to manage fluorescence features of doped bulk and nanosized materials by variation of excitation temporal characteristics. It can be done by variation of pulse length, pulse repetition rate or implementing different types excitation pulses train. Laser-like effect was revealed in Yb,Tm:YF_3 single crystal under excitation by rectangular pulse train at 930-980 nm spectral domain. $^3H_4$-$^3F_4$ transitions of Tm^{3+} ions possess stimulated emission radiation features which appear above certain threshold excitation power and characterized by jump-like evaluation of fluorescence kinetics. This was also observed for higher lying ^3P_j, $^1I_6$, $^1D_2$, $^1G_4$ manifolds obviously populated due to excited-state transitions and up-conversion process in Yb,Tm:YF3. These results open an opportunity of elaborating high effective upconversionally pumped microchip solid-state lasers.

Acknowledgements

This work was funded by the subsidy allocated to Kazan Federal University for the project part of the state assignment in the sphere of scientific activities.

References

[1] Wang F and Liu X 2014 *Accounts of Chemical Research* **47** 1378
[2] Ninjbadgar T, Garnweitner G, Borger A, Goldenberg L M, Sakhno O V and Stumpe J 2009, *Adv. Funct. Mater.* **19** 1819
[3] Huang X, Han S, Huang W and Liu X 2013 *Chem. Soc. Rev.* **42** 173-201
[4] Auzel F 2004 *Chem. Rev.* **104** 139
[5] Antipenko B M, Mak A A, Raba O B, Seranjan K B, and Uvarova T B 1983 *Sov. J. Quantum Electron.* **13** 558.
[6] Ostermayer F W, Jr., Van der Zeil J P, Marcos H M, Van Uitert L G, and Geusic J E 1971 *Phys.Rev.B* **3** 2698
[7] Pavlov V V, Kazakov B N, Lovchev A V 2014 *Pis'ma v ZhETF* **100** 13
[8] To be published in *J. of Lum.*

TERS microscopy as a probe for visualizing the orientation of chromophores embedded in the glassy polymers

K L Nefedyeva, S S Kharintsev, A I Fishman, M Kh Salakhov

Kazan Federal University, 420008, Kremlevskaja str., 18, Kazan, Russian Federation

E-mail: nefedieva_ksu@mail.ru

Abstract. Photo-induced effects on azobenzene-functionalized polymer thin films under exposure to strongly focused higher laser modes are investigated. Tip-enhanced Raman scattering microscopy provides a possibility for probing and visualizing the orientation of anisotropic chromophores embedded in the glassy environment at room temperature.

1. Introduction

Azobenzene containing polymer thin films are technologically important structures. The appropriate electron-donor and electron-acceptor substitutes result in large polarizability and intrinsic dipole moment of azochromophores [1]. The ordering of dipoles produces a high optical nonlinearity and therefore the second harmonic generation (SHG) is observed in these patterns [2,3]. This is one of the research directions of azobenzene-based matter [4,5].

On the other hand thin films with azochromophores possess photomechanical properties [6,7]. This allows the creation of different structures on the film surface by the laser irradiation. For instance interesting application is the photo-assisted fabrication of the surface relief gratings [8,9]. Such structures can be used as diffractive optical elements [6]. Moreover, the periodical orientation of dipoles in such gratings is one of the approaches of increasing the SHG efficiency.

In our work we concentrate on the molecular motion induced by strongly focused laser beam with linear and radial polarization. The observed structures are analyzed with the use of tip enhanced Raman scattering (TERS) to estimate the presence of optical anisotropy.

2. Experiment and discussion

The movement of azopolymer is strongly dependent on the orientation of the incident light. We first consider the interaction of azopolymer with the linear polarized laser beam. It is assumed that the incident light is tightly focused by using the high aperture objective. The incident light firstly leads to the orientation of induced dipole moments (figure 1). Secondly the inhomogeneous optical field results in the arising of the gradient force [6]:

$$\vec{f} = \frac{1}{4}\varepsilon_0\chi'\exp\left(-\alpha z\right)\frac{\partial I(x,y)}{\partial x}\vec{x}_0,$$

where ε_0 is a permittivity of vacuum, χ' is the real part of the polymer medium susceptibility, α is the absorption coefficient of light, $I(x,y)$ is the intensity distribution of the Gaussian beam at the focal plane and $\vec{x}_0$ is the unit vector along the x axis direction. Thus, the azobenzene chromophores experience the force only in the polarization direction along the component of the optical-field gradient, and the force is vanishing when the polarization is orthogonal to the optical field. The moving of the polymer mass occurs only in the polarization direction and leads to the appearing of the transverse profile on the film surface.

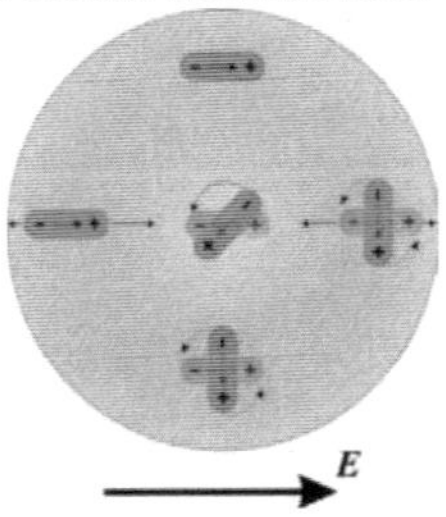

Figure 1. The orientation and displacement of the dipoles in the Gaussian beam (top view).

We have confirmed this experimentally. The object of our investigation is 4-amino-4'-nitroazobenzene covalently connected to the main chain of the polymer matrix (CFAO) (fig. 2a) [10]. Amino group is the donor of electrons and nitro group is the electrons acceptor. CFAO was strongly diluted in cyclohexanone. The films were casted on slide cover glass at 5000 to 12000 rpm. Thereafter the films were annealed during 1 hour at a temperature of 100°C for removing solvent remains. A film thickness was estimated to be ~70 nm.

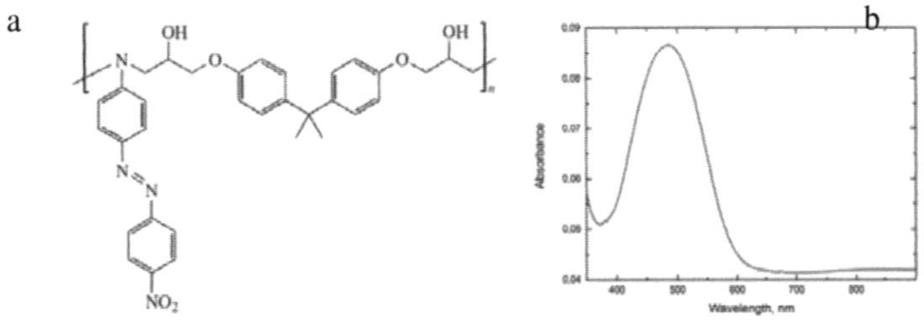

Figure 2. The structure of CFAO (a) and its absorption spectrum (b).

In order to produce the photoinduced surface deformation the polymer films with azobenzene were irradiated by the linearly polarized laser beam with the 633.8 nm wavelength which is far from the maximum of the absorption (fig. 2b). The exposure time was 100 s and the intensity at the center of a laser spot was about a few kW/cm^2. A high numerical aperture (NA=1.4) oil-immersion lens (100×) squeezes a laser beam into sizes less than diffraction-limited laser spot thus providing information on high spatial frequencies.

The linearly polarized light engraves a dip and pitch on the CFAO surface film at the centre and rims of the laser spot respectively. The topography and the cross section of the structure are shown in figure 3. A depth of the dip is approximately 3 nm and the diameter is about 1.5 μm. Maximal surface deformation was observed along the laser polarization. Thus, the experimental data are in agreement with the theoretical model.

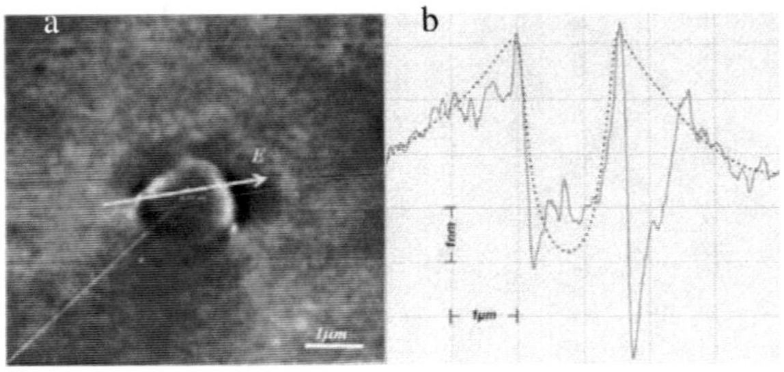

Figure 3. AFM of polymer surface after linear polarized laser radiation (a) and its cross section along polarization direction (b).

Importantly, the deformation profile is primarily dependent on the light polarization. In particular, the radial polarization of the highly focused light beam permits the longitudinal field to be as great as the transverse field. For this purpose we implemented a four sectional radial polarizer consisting of four oriented half-waveplates (fig. 4).

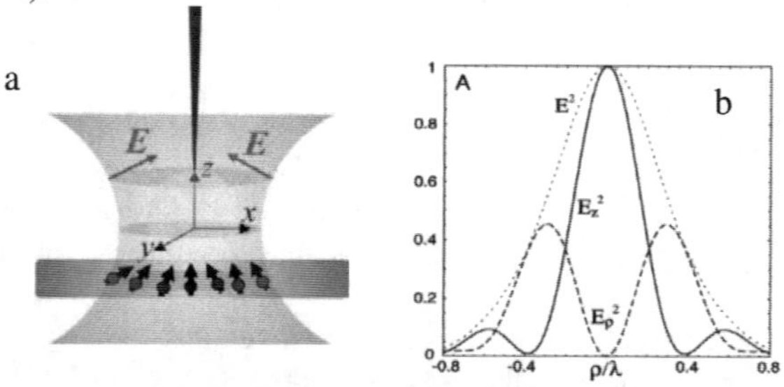

Figure 4. Tightly focused radial polarized laser beam, longitudinal component enhanced by optical antenna (a) and optical field distribution (b). $\rho = \sqrt{x^2 + y^2}$.

The use of the optical antenna results in the field enhancement by several orders of magnitude. The tip squeezes the far field over the extent of several nanometers and, thus, sub-wavelength patterns occur. In figure 5 the AFM image of the film surface after illumination by optical longitudinal field is shown. The protrusion with a height of 10 nm and a diameter of 100 nm appears at the centre of the laser spot.

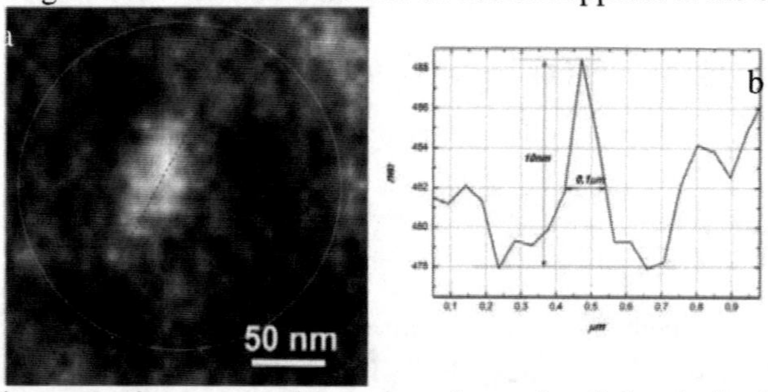

Figure 5. AFM of polymer surface exposed to a tip-enhanced radial polarized light (a) and its cross section (b). The area of the laser beam placing is marked by dotted circle.

To analyze the molecular orientation we have compared Raman spectra of the CFAO film before and after corona discharge poling using the 632.8 nm excitation. We have found no difference between them at moderate intensities of the laser light. At higher intensities we have observed degradation and photobleaching of the CFAO film. This is evidenced by the irreversible appearance of a peak at 1130 cm^{-1}, as shown in figure 6.

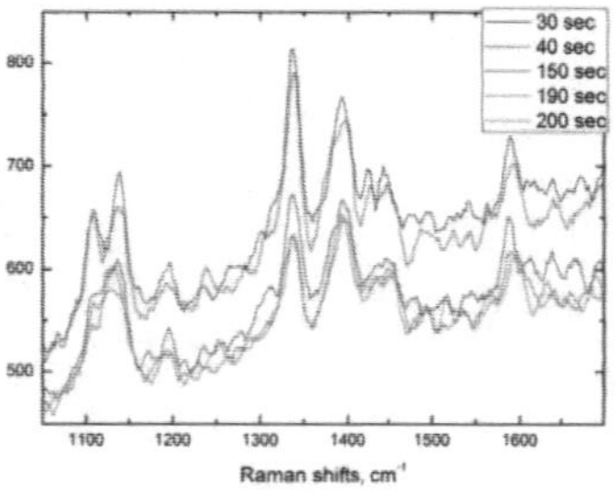

Figure 6. Raman spectra of the CFAO films with different exposure times.

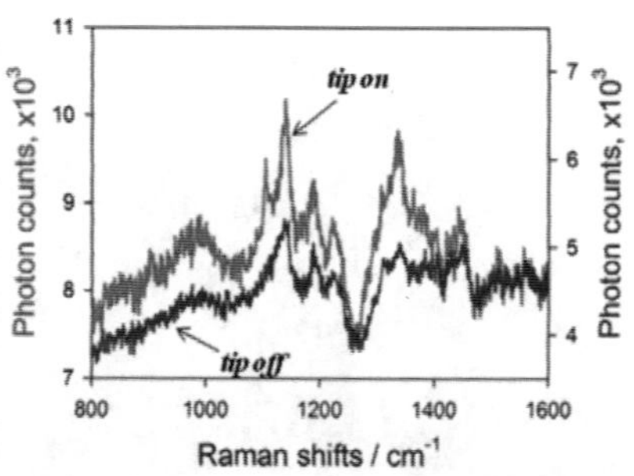

Figure 7. Near- and far-field Raman spectra of the CFAO film with the 632.8 nm excitation.

The absorption of chromophores depends on the polarization of incident light. Figure 7 shows a Raman spectrum of the original 70 nm thick film, that has been recorded with 632.8 nm radially polarized light for 100 s. TERS method allows one to probe the longitudinal anisotropy with a sub-wavelength spatial resolution, whereas the transverse anisotropy can be visualized with the diffraction-limited resolution.

The optical antenna can convert the far field into localized modes due to the plasmonic effect [11]. A hot spot beneath the tip apex is excited with longitudinal fields, which are parallel to the long axis of the conical antenna. Tip-enhanced Raman scattering comes from chromophores whose main polarizability components (α_{xx} or α_{zz}) are oriented towards the longitudinal field. If one of the directions coincides with that of a dipole moment then we observe a giant absorption and photodegradation appears. The enhancement of a peak at 1110 cm^{-1} provides evidence that chromophores are mainly aligned along the film surface rather than inward the bulk. As the chromophore gets oriented along the longitudinal field, it is inevitably photobleached and destroyed. Figure 8 demonstrates TERS maps of the oriented thin film at 1130 cm^{-1} and 1400 cm^{-1}. The presence of the peak at 1130 cm^{-1} (fig.8 a) is a result of chromophore orientation along the longitudinal field and, therefore, the molecule has been photobleached. On the other hand, 1400 cm^{-1} peak (fig.8 b) indicates that the N=N bond for of out-of-plane anchored chromophores is conserved.

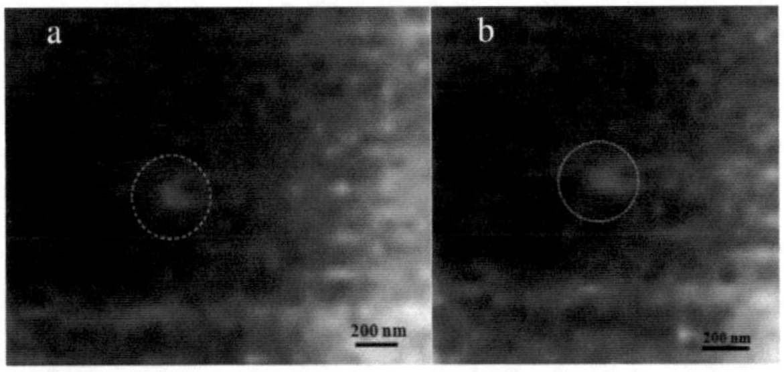

Figure 8. The Raman mapping of the CFAO thin films at 1130 cm^{-1} (a) and 1400 cm^{-1} (b) The area of the laser beam placing is marked by dotted circles.

3. Conclusion

The light-driven surface deformations of azobenzene chromophore containing polymer thin films were achieved by using linear and radial polarized incident light. It has been obtained TERS spectra of the thin films and found the feature for aligning chromophores in the thin film.

Acknowledgements

This work was funded by the subsidy allocated to Kazan Federal University for the state assignment in the sphere of scientific activities. This work was financially supported by the Russian Foundation for Basic Research (No. 13-02-00758 A). This work was done using equipment of Federal Center of Shared Equipment of Kazan Federal University.

References

[1] Mahimwalla Z, Yager K G, Mamiya J, Shishido A, Priimagi A, Barrett C J 2012 *Polym. Bull.* **69** 967
[2] Singer K D, Sohn J E and Lalama S J 1986 *Appl. Phys. Lett.* **49** 248
[3] Wang Y, Tai O Y-H and Wang C H 2005 *J. Chem. Phys* **123** 164704
[4] Han L, Jiang Y, Li W, Li Y and Liao J 2008 *Materials Letters***62** 1495
[5] Rodriguez F J, Sanchez C, Villacampa B, Alcala R, Cases R, Millaruelo M and Oriol L 2004 *Polymer* **45** 2341
[6] Bian S, Williams J M, Kim D Y, Li L, Balasubramanian S, Kumar J and Tripathy S 1999 J. *Appl. Phys.***86** 4498

[7] Ishitobi H, Nakamura I, Kobayashi T, Hayazawa N, Sekkat Z, Kawata S and Inouye Y 2014 *ACS Photonics* **1** 190
[8] Schaller R D, Saykally R J, Shen Y R and Lagugne-Labarthet F 2003 *Optics Letters* **28** 1296
[9] Fiorini C, Prudhomme N, Veyrac G, Maurin I, Raimond P and Nunzi J-M 2000 *Synthetic Metals* **115** 121
[10] Nikonorova N A, Balakina M Yu, Fominykh O D, Pudovkin M S, Vakhonina T A, Diaz-Calleja and Vakimansky A V2012 *Chem. Phys. Lett.* **552** 114
[11] Novotny L and Hecht B 2006 *Principles of Nano-Optics* (Cambridge: University Press)

Determination of the size of. nanoparticles in photonic nanostructures from AFM images

A A Akhmadeev and M Kh Salakhov

Kazan Federal University, 420008, Kremlevskaja str., 18, Kazan, Russian Federation

E-mail: Albert.Akhmadeev@kpfu.ru

Abstract. In this paper we propose a relatively simple algorithm for determining the size of ellipsoidal particles on atomic force microscopy (AFM) image. The proposed algorithm frees researchers from the "manual" measuring of the size of a large number of particles. This algorithm is based on a three-dimensional approximation of the segmented images by ellipsoids and it has several advantages over similar methods. One of important advantages is the ability to determine the parameters of particles the surface of which is not completely visible on the image. Proposed method has been tested on simulated images as well as on the experimentally obtained AFM images of silica particles. Especially the method is very useful for the study of opal-like photonic crystals, in which the particles are packed tightly.

1. Introduction

Recently, advanced materials consisting of micro- and nanoparticles have received considerable attention of researchers. These materials are widely used in various areas including applied physics, chemistry and biology. Such materials include photonic nanostructures and photonic crystals [1]. The structure of photonic crystal is characterized by periodic distribution of refraction index in the spatial directions so far it has its own scheme of electromagnetic states and the photonic band gaps in the spectrum. With the advent of devices such as a scanning electron microscope (SEM) and atomic force microscope (AFM), it was possible to visualize the surface of nanostructured materials and investigate their properties. The particle size is one of basic characteristics of these materials because it influences a large variety of physical, electronic, and optical properties. Also, the size of the particles of photonic crystal could be crucial for the manifestation of a new quantumelectrodynamical effect predicted in earlier works [2–4]. Thus, accurate determination of the size is an important task [5–8]. In this paper we developed a relatively simple algorithm that allows determining the size of a large number of ellipsoidal particles on AFM images.

2. Algorithm for determining the particle size

Many existing methods for determining the size of spherical particles on microscopy images include two basic steps:

1) Transformation of the original image (usually obtained by SEM) to monochrome using special filters (e.g., Canny [9], Sobel [10] filters). Result of this transformation is an image containing only the outlines of particles.

2) Recognition of the outlines by Hough transform [11], or analogous methods. Thus, the outline of each particle is mapped to a circle whose diameter is approximately equal to the particle size.

The considered approach is not free from drawbacks. In the case of some of the particles on the image are not completely visible (e.g., the particles are covered by other particles) their observable outline will not match the actual boundaries of the particle thus leading to incorrect results.

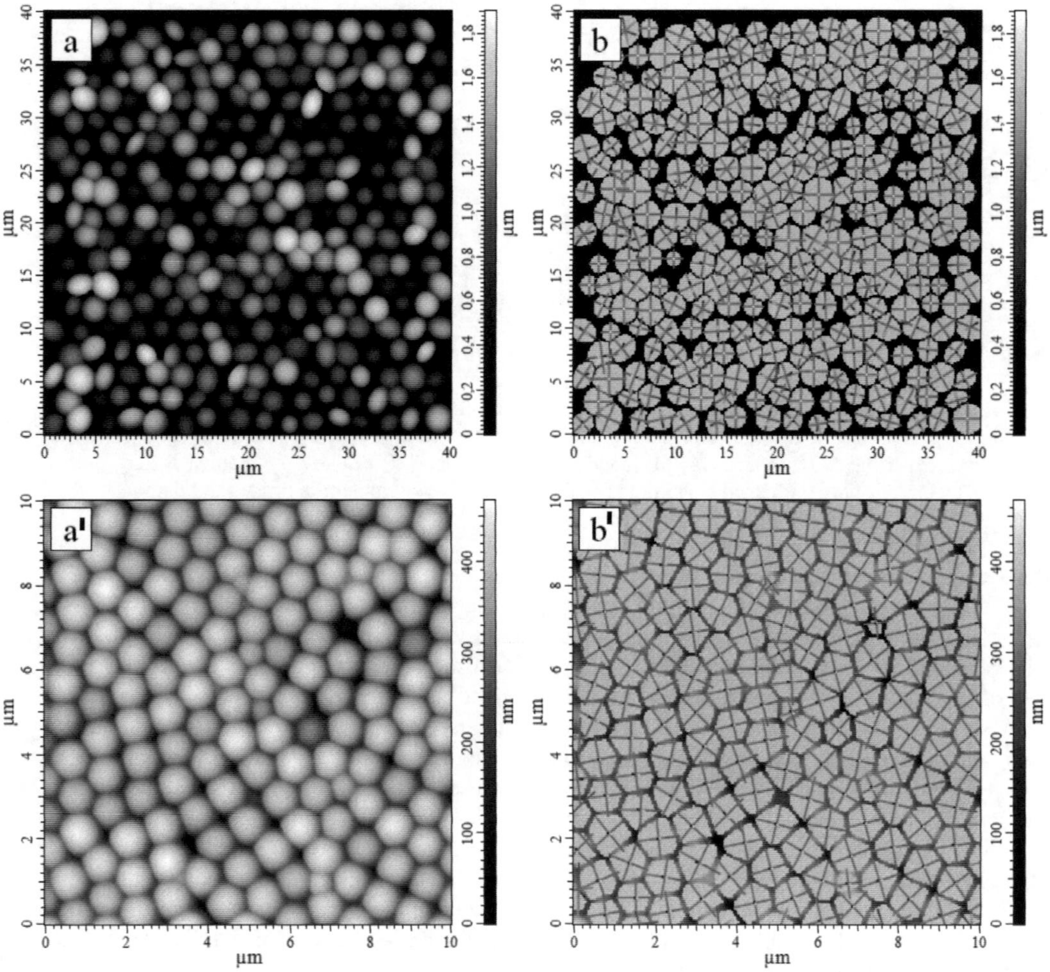

Figure 1. Processing of AFM image on the basis of three-dimensional approximation by quadric surfaces. The simulated and real AFM image of particles (a) and (a') correspondingly and its segmentation into approximated ellipsoids (b) and (b').

The technique proposed in this paper does not have such a drawback, since it utilizes an important advantage of AFM – 3D-visualization of sample surface. Thus, scanning the sample surface by atomic force microscopy provides an image that represents two-dimensional array of heights $Z_{i,j}$, corresponding to the sample topography. We propose determining the particle sizes on these images via the geometric curvature, instead of particle boundaries. To this end the set of points of $Z_{i,j}$ are approximated by quadric surfaces (sphere and ellipsoid are special cases), described by the equation:

$$a_{11}x^2 + a_{22}y^2 + a_{33}z^2 + 2a_{12}xy + 2a_{13}xz + 2a_{23}yz + b_1x + b_2y + b_3z + c = 0 \quad (1)$$

The origin of coordinate system is chosen so that $c \neq 0$. Thus, the approximation is reduced to finding the nine independent parameters (a_{ij}, b_j) in the equation (1). Using the method of least squares (OLS) these parameters can be found by solving a system of linear equations (case of linear regression).

But primarily, the input image should be segmented, where each segment represents a set of points in the array $Z_{i,j}$. At the same time, these segments must conform to the individual particles. Since the particles have a convex surface, then the following conditions are satisfied:

$$\frac{\partial^2 Z}{\partial x^2} < 0, \quad \frac{\partial^2 Z}{\partial y^2} < 0 \quad (2)$$

Because the particle surface on AFM image is represented as an array $Z_{i,j}$ instead of a continuous function, derivatives in (2) are replaced by the differences:

$$Z_{i-1,j} - 2Z_{i,j} + Z_{i+1,j} < 0, \quad Z_{i,j-1} - 2Z_{i,j} + Z_{i,j+1} < 0 \quad (3)$$

All points (i,j) satisfying the condition (3) are distributed on separate groups, each group corresponds to a single particle. So the proposed method consists of the following steps:

1) Obtaining AFM images of micro- or nanoparticles (as an array).
2) Segmentation of the array into groups of points satisfying the condition (3).
3) Approximation of each group of points by quadric surface using OLS.
4) Finding the transformation to a canonical form for a quadric surface.

As a result of this procedure the following parameters are determined for each particle on the AFM image with the shape close to an ellipsoid: coordinates of the center, length of the semiaxes and their orientation in space.

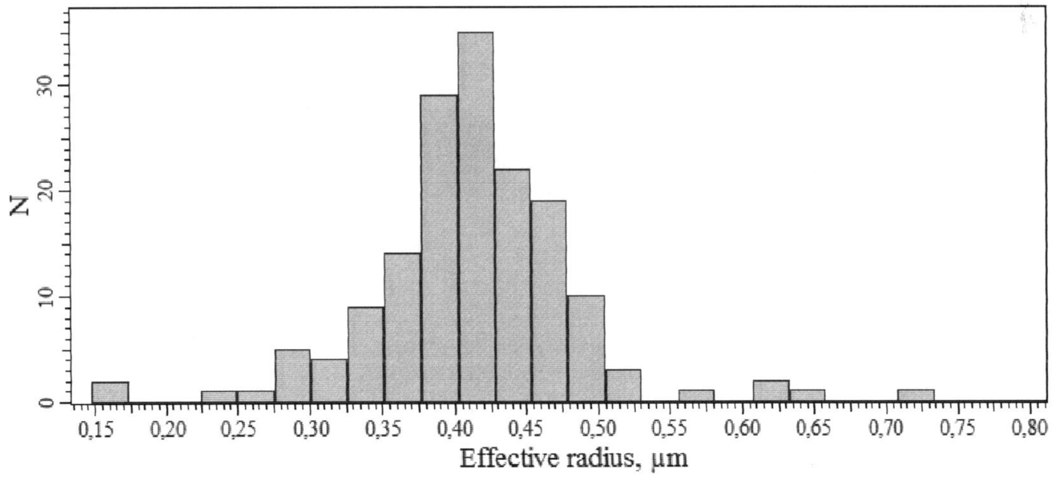

Figure 2. Distribution of effective radius of silica particles.

3. Verification of the algorithm on simulated and experimental images

To verify the described method, images containing spheres and ellipsoids have been simulated (figure 1 a), each particle has a predetermined parameters. Filled areas in figure 1 (b) correspond to groups of points in the array $Z_{i,j}$, which satisfy the condition (3), the approximation of which yielded the output parameters: the length of the semi-axes, coordinates of the centers and three orthonormal vectors showing the orientation of the axes of the ellipsoid. Output parameters with high precision coincide with the input parameters of the ellipsoids; the difference in particle size does not exceed 10^{-8} nm. Figure 1 (b) also shows the orientation of each ellipsoid, obtained from the image processing by our method.

As can be seen from figure 1, this technique is able to handle with particles which are covered by other particles, and particles that cling together. By applying this algorithm to real AFM image of silica particles (SiO_2) the sizes and orientations of the particles were determined (figure 1 a' and b'). The particles of SiO_2 were obtained by hydrolysis of orthosilicic acid (tetraethoxysilane, TEOS), under the action of a catalyst (Stober method [12]). Such particles work as a building material for the preparation of colloidal photonic crystals [13, 14]. Distribution of effective radius of these particles is shown in figure 2, where the effective radius is the geometric mean of semiaxes of the ellipsoid ($R = \sqrt[3]{abc}$).

4. Conclusions

Thus, we have proposed a relatively simple algorithm that allows rapid determining of the size and spatial orientation of the particles whose shape is close to an ellipsoid. The results show that the algorithm allows determining the parameters of particles that are not completely visible on the image.

Acknowledgements

This work was funded by the subsidy allocated to Kazan Federal University for the state assignment in the sphere of scientific activities.

References

[1] Joannopoulos J D, Johnson S G, Winn J N and Meade R D 2008 *Photonic crystals: molding the ow of light* (Princeton: Princeton University Press) p 286
[2] Gainutdinov R K, Khamadeev M A and Salakhov M K 2012 *Phys. Rev. A* **85** 053836
[3] Gainutdinov R Kh, Khamadeev M A and Salakhov M Kh 2013 *Bull. Russ. Acad. Sci. Phys.* **77** 1440–1443
[4] Gainutdinov R Kh, Khamadeev M A and Salakhov M Kh 2013 *J. Phys.: Conf. Ser.* **478** 012017
[5] Sedlar J, Zitova B, Kopecek J, Flusser J, Todorciuc T and Kratochvilova I 2013 *J. Nanopart. Res.* **15** 1842
[6] Garnaes J 2011 *Meas. Sci. Technol.* **22** 094001
[7] Samusev K B, Yushin G N, Rybin M V and Limonov M F 2008 *Physics of the Solid State* **50** 1280–1286
[8] Chuklanov A P, Ziganshina S A and Bukharaev A A 2006 *Surf. Interface Anal.* **38** 679–681
[9] Canny J 1986 *IEEE Trans. Pattern Anal. Mach. Intell.* **8** 679–698
[10] Sobel I 1990 *An Isotropic 3x3 Gradient Operator, Machine Vision for Three-Dimensional Scenes ed H Freeman* (New York: Academic Press) pp 376–379
[11] Gonzalez R and Woods R 1992 *Digital image processing* (Boston: Addison-Wesley) p 716
[12] Stober W, Fink A and Bohn E 1968 *J. Colloid Interface Sci.* **26** 62–69
[13] Akhmadeev A A, Sarandaev E V and Salakhov M Kh 2013 *J. Phys.: Conf. Ser.* **461** 012022
[14] Koryukin A V, Akhmadeev A A and Salakhov M Kh 2013 *J. Phys.: Conf. Ser.* **478** 012013

Journal of Physics: Conference Series **560** (2014) 012006 doi:10.1088/1742-6596/560/1/012006

Self-energy function of quantum-dot states and resonance fluorescence

R Kh Gainutdinov, M A Khamadeev, M R Mohebbifar, A A Mutygullina

Department of Physics, Kazan Federal University,
18 Kremlevskaya St, Kazan 420008, Russia

E-mail: Renat.Gainutdinov@kpfu.ru

Abstract. We investigate the energy dependence of self-energy functions of quantum-dot states that describe the interaction of quantum dots with vacuum. It is shown that the variation of these self-energy functions can be significant and this can have a significant effect on processes of quantum interference from quantum dots.

1. Introduction

Resonance fluorescence from artificial atoms and quantum dots [1, 2] is now a major field of research. As was predicted by Mollow [3] and confirmed by a number of experiments [4–6], at relatively high laser intensities the fluorescence spectrum should consist of a symmetric triplet whose side components are separated from the central peak by distances equal to the generalized Rabi frequency. However, the recent experimental observations of the resonance fluorescence from the artificial atom and quantum dots [1, 2] showed an essential discrepancy with the predictions of quantum optics for two-level atomic systems. The observations of the resonance fluorescence from self-assembled quantum dots, being atom-like two-level emitters, showed [2] not only the asymmetry which for some detuning becomes very significant but also surprising sideband linewidths: there is a linear-linear zoom-in on Mollow sidebands at some laser powers. The discrepancy between the experiment and the theory could originate from the fact that the ordinary theory of the resonance fluorescence does not take into account the fact that the effect of the vacuum modes on the spectrum of resonance fluorescence may be significant. The fact that the interaction of emitters with the vacuum modes can be very significantly modified in nanomaterials has been demonstrated in [7]. In particular this should have a significant effect on the Lamb shift of the components of the Mollow triplet. In Refs. [8] it has been shown that under some conditions on the self-energy function of emitter states the contributions from the vacuum modes cannot be treated as small corrections. The aim of the present paper is to show that the above conditions can be realized in processes of resonance fluorescence from quantum dots.

2. Vacuum modes and resonance fluorescence from quantum dots

The strong interaction of an atom with the laser field resonant with an atomic transition is successfully treated within the quantum optics formalism [9,10]. When the laser mode frequency

Journal of Physics: Conference Series **560** (2014) 012006 doi:10.1088/1742-6596/560/1/012006

ω_L is close to the frequency $\omega_R = \omega_e - \omega_g$ of the transition between two bound states $|e\rangle$ and $|g\rangle$ the atom-laser interaction is strong. In quantum optics this interaction is usually described by using the two-level model and the rotating wave approximation (RWA). In this approximation the laser-dressed states are the eigenstates of the system Hamiltonian H_{RWA} [10] and are given by

$$|+, n\rangle = \cos\theta_n|e, n\rangle + \sin\theta_n|g, n+1\rangle; \quad |-, n\rangle = -\sin\theta_n|e, n\rangle + \cos\theta_n|g, n+1\rangle. \tag{1}$$

Here $|e, n\rangle$ ($|g, n\rangle$) denotes the state of the combined laser-atom system containing the atom in the bare state $|e\rangle$ ($|g\rangle$) and n photons in the laser mode, and θ_n is the mixing angle defined by $\tan(2\theta_n) = -\Omega_n/\Delta$ with Ω_n and Δ being the Rabi frequency $\Omega_n = 2g_L\sqrt{n+1}$ and detuning, respectively. The energies of the dressed states are given by $E_{\pm,n} = (n+1/2)\omega_L + (\omega_e - \omega_g)/2 \pm \Omega_R^{(n)}/2$ with $\Omega_R^{(n)}$ being the generalized Rabi frequency $\Omega_R^{(n)} = \sqrt{\Omega_n^2 + \Delta^2}$. The states dressed by the interaction of the atom with the resonance laser field are not dressed by its interaction with the vacuum modes, and in our investigation of the problem we do not assume a priori that this interaction is weak in any case and use nonperturbative methods of its description based on the the generalized dynamical equation (GDE), which in [11] has been derived as a direct consequence of the first principles of quantum physics. Being equivalent to the Schrödinger equation in the case when the interaction in a quantum system is instantaneous, GDE allows one to extend quantum dynamics to the case of nonlocal-in-time interactions. This equation has been proven to provide a new insight of many problems in nuclear physics [12–15], atomic physics [16–20] and quantum optics [8,21]. Within this approach the QED bound state problem is solved by using the formalism of the self-energy function $C_n(z)$ (see for details Refs. [22]), that describes the interaction of the particles in the state $|n\rangle$ with the vacuum. The self-energy QED corrections to the energies $E_n^{(0)}$ of the bare states are determined by the equation $z - E_n^{(0)} - C_n(z) = 0$ where $C_n(z)$ in turn are determined by GDE. In the first approximation the solution of this equation is $E_n \equiv E_n^{(0)} + C_n^{(0)}(E_n^{(0)}) \equiv E_n^{(0)} + \Delta E_n^L - \frac{i}{2}\Gamma_n$, where E_n is the energy of the real dressed state, E_n^L is the Lamb shift, Γ_n is the natural level width, and $C_n^{(0)}(z)$ is determined by the equation

$$\frac{dC_n^{(0)}(z)}{dz} = -\langle n|H_I\tilde{G}_0^2(z)H_I|n\rangle, \tag{2}$$

with H_I being the QED Hamiltonian. For this approximation to be valid, probability amplitudes of the nonradiative transitions between the states with the same total angular momentum J, its projection J_z and parity must be much smaller than the energy distance between these states. In addition the variation of the function $C_n(z)$ with the energy must be negligible on the energy interval of order of the value of the Lamb shift. This takes place in the case when the atom is in free space. In fact, in this case the energy distance between the states with the same J, J_z and parity are of order of $\alpha^2 m_e$, while the probability amplitudes of nonradiative transitions between them are of order of $\alpha^5 m_e$, and the self-energy function varies significantly with the energy only on energy intervals of order m_e. The situation is dramatically changed in the case of an emitter subject to resonance laser field because the energy distance between dressed states $|+, n\rangle$ and $|-, n\rangle$ having the same J, J_z and parity equals the Rabi frequency. As a result the nonradiative transitions between these states begin to play an important role. At leading order the probability amplitudes $\langle\pm, n|\Sigma(z)|\mp, n\rangle$ of these transitions are given by the formula [8] $\langle\pm, n|\Sigma(z)|\mp, n\rangle = \cos\theta\sin\theta(C_g^{(0)}(z) - C_e^{(0)}(z))$, where $C_g^{(0)}(z)$ and $C_e^{(0)}(z)$ are the self-energy functions of the bare states $|g\rangle$ and $|e\rangle$ respectively. At the same time, in the case of an atom being in free space the energy interval of order Ω_R is too small for the variation with energy of $\langle\pm, n|\Sigma(z)|\mp, n\rangle$ to be noticeable. In this case the energy dependence of $\langle\pm, n|\Sigma(z)|\mp, n\rangle$ can be neglected, and they can be regarded as matrix elements of an additional term in the Hamiltonian describing the interaction of the atom with resonance laser field. Adding such a term to the RWA Hamiltonian will lead only to a change in the mixing angle θ_n, and, as a consequence, will not give rise to asymmetry of the Mollow spectrum. However, as we show below, in the case when a quantum dot acts as an emitter the variation of the self-energy functions $C_g(z)$ and $C_e(z)$ on an energy interval compared to the energy difference between the dressed states $|+,n\rangle$ and $|-,n\rangle$ can be very significant, and hence the vacuum modes are involved into the strong atom-laser interaction.

3. Self-energy functions of the quantum-dot states

Let us consider the model of a quantum dot interacting with two reservoirs in which the Hamiltonian of a system consisting of quantum dot and leads is described by $H = H_{QD} + H_T + H_{lead}$, where H_{QD} is the Hamiltonian describing the quantum dot, H_{lead} is the lead Hamiltonian, H_T is the Hamiltonian which describes tunneling processes between dots and leads. The single-particle level spacing in the dot is assumed to be larger than any other energy scale (temperature, Coulomb interaction, and transport voltage) so that only one energy level needs to be taken into account. The quantum dot can be described by the single-level Anderson impurity model, in which H_{QD} is of the form $H_{QD} = \sum\limits_{\sigma=\uparrow,\downarrow} Ed_\sigma^\dagger d_\sigma + Un_\uparrow n_\downarrow$, where the creation (annihilation) operator for an electron with spin σ on the dot is given by $d_\sigma^\dagger$ and $d^\dagger$, and $n_\sigma = d_\sigma^\dagger d_\sigma$ is the corresponding number operator. The on-site repulsion U describes the energy cost for double occupation (when we have two electrons in quantum dot) and stems from Coulomb interaction. The Hamiltonian H_T is given by $H_T = \sum\limits_{s,k,\sigma} V_s C_{s,k,\sigma}^\dagger d_\sigma + H.C.$, where V_s is the momentum- and spin-independent tunnel matrix element, $C_{s,k,\sigma}^\dagger (C_{s,k,\sigma})$ is creation (annihilation) operators for electrons with spin σ and momentum k in lead and $s = L, R$. The Hamiltonian H_{lead} is given by $H_{lead} = \sum\limits_{s,k,\sigma} E_{s,k} C_{s,k,\sigma}^\dagger C_{s,k,\sigma}$. The chemical potentials of the two leads differ by the applied bias $\mu_L - \mu_R = -eV$.

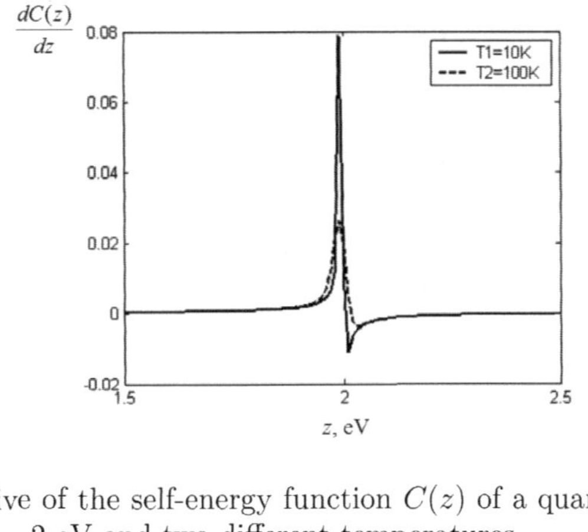

Figure 1. The derivative of the self-energy function $C(z)$ of a quantum-dot state for $R = 0.1$ eV, $U = 30R$, $\mu_L = \mu_R = 2$ eV and two different temperatures

We assume that the density of states ρ_s in the leads is constant for transport and determines the tunnel coupling strength R_s as $R_s = 2\pi\rho_s \mid V_s \mid^2$ and $R = R_L + R_R$. The Hilbert space for the quantum dot is spanned by the states $|0\rangle$ for an empty dot, $|\sigma\rangle$ for a singly occupied dot with spin $\sigma = \uparrow, \downarrow$ and $|d\rangle$ for a doubly occupied dot. The corresponding energies are E_0, E_σ, and E_d. By generalizing the equations derived in [23] for the self-energy function of quantum dot we get

$$C(z) = -\sum_{s} \frac{R_s}{2\pi} \int d\omega \left(\frac{f_s(\omega)}{\omega - z} + \frac{f_s(\omega)}{z + U - \omega} \right), \tag{3}$$

where $f_s(\omega) = (1 + \exp(\frac{\omega - \mu_s}{K_B T}))^{-1}$ is the Fermi distribution function. As it follows from the results of calculations presented in Fig. 1 the variation of the self-energy function is very significant in the vicinity of the point $z = 2$ eV.

4. Conclusions

In this paper we have investigated the self-energy functions of quantum dots, in order to understand the effect of vacuum modes on the process of resonance fluorescence from quantum dots. We have shown that the variation of these functions on the energy interval compared to the energy difference between the dressed states $|+, n\rangle$ and $|-, n\rangle$ is very significant. This means that the vacuum modes have a significant effect on the resonance fluorescence from quantum dots, and the development of a theory that could take them into account in consistent way could remove the discrepancy between experiment and the theoretical predictions.

Acknowledgments

This work was funded by the subsidy allocated to Kazan Federal University for the state assignment in the sphere of scientific activities.

References

[1] Astafiev O 2010 *Science* **327** 840
[2] Vamivakas A N, Zhao Y, Lu C-Y, Atatüre M 2009 *Nature Phys.* **327** 840
[3] Mollow B R 1969 *Phys. Rev.* **188** 1969
[4] Wu F Y, Grove R E and Ezekiel S 1975 *Phys. Rev. Lett.* **35** 1426
[5] Shuda F, Stroud Jr C R and Hercher M 1974 *J. Phys.* B **198** L7
[6] Grove R E, Wu F Y and Ezekiel S 1977 *Phys. Rev.* A **15** 227
[7] Gainutdinov R Kh, Khamadeev M A and Salakhov M Kh 2012 *Phys. Rev.* A **85** 053836.
[8] Gainutdinov R Kh and Mutygullina A A 2009 *Physica Scripta* **T135** 014021
[9] Mandel L and Wolf E 1995 *Optical Coherence and Quantum Optics* (Cambridge: Cambridge University Press)
[10] Scully M and Zubairy M 1997 *Quantum Optics* (Cambridge: Cambridge University Press)
[11] Gainutdinov R Kh 1999 *J. Phys.* A **32** 5657
[12] Gainutdinov R Kh, Moutygoullina A A 1997 *Physics of Atomic Nuclei* **60** 841
[13] Gainutdinov R Kh, Moutygoullina A A 1999 *Physics of Atomic Nuclei* **62** 1905
[14] Gainutdinov R Kh, Moutygoullina A A 2002 *Physics of Atomic Nuclei* **65** 1421
[15] Gainutdinov R Kh, Mutygullina A A, Khamadeev M A, Petrova A S 2013 *J. Phys.: Conf. Ser.* **478** 012019
[16] Gainutdinov R Kh, Khamadeev M A and Salakhov M Kh 2013 *J. Phys.: Conf. Ser.* **478** 012017
[17] Gainutdinov R Kh, Mutygullina A A, Scheid W 2001 *Proc. of SPIE* **4605** 214
[18] Gainutdinov R Kh, Mutygullina A A, Scheid W 2001 *Proc. of SPIE* **4605** 222
[19] Gainutdinov R Kh 1995 *ZETP* **108** 1600
[20] Gainutdinov R Kh 1986 *Optika i Spektroskopiya* **60** 890
[21] Gainutdinov R Kh, Mutygullina A A 2013 *J. Phys.: Conf. Ser.* **478** 012018
[22] Gainutdinov R Kh 1984 *Ukrainskii fizicheskii zhurnal* **29** 805
[23] Splettstoesser J, Governale M and König J 2012 *Phys. Rev.* B **86** 035432

Journal of Physics: Conference Series **560** (2014) 012007 doi:10.1088/1742-6596/560/1/012007

Singular behavior of the photon density of states and the self-energy function of an electron in photonic crystal

R Kh Gainutdinov, M A Khamadeev, D N Avramenko, K A Ziyatdinova, M Kh Salakhov

Department of Physics, Kazan Federal University,
18 Kremlevskaya St, Kazan 420008, Russia

E-mail: Renat.Gainutdinov@kpfu.ru

Abstract. We show that Van Hove singularities in the behavior of the photon density of states in a photonic crystal results in a singular behavior of the self-energy functions of atomic states that describe the interaction of the atom with vacuum. This is shown to have a significant effect on processes of quantum interference from atoms in the PC medium.

1. Introduction

There has been a growing interest in the study of photonic crystals (PC) which are periodic systems consisting of dielectric materials with different refractive indices. The variation of the photon density of states (DOS) being a result of a modification of electromagnetic field in PCs leads to quantum effects, including the coherent control of the spontaneous emission [1], the appearance of photon-atom bound states [2], the non-Markovian character of radiative decay [3], enhanced quantum interference effects [4] and the localization of superradiance near the photonic band edge [5]. In Ref. [6] it has been shown that a strong modification of the interaction of a charged particle with its own radiation field in photonic crystals results in the fact that in the PC medium the electron mass changes its value. This effect is important not only from the fundamental point of view but also for applications. It opens new ways of using PCs [7] that in contrast to the ordinary ways are not based on the band gap effect. The modification of the self-interaction of electrons affect not only the electron mass but also the character of the interaction of photons with atoms in PC's. The aim of the present paper is to show that this can have a significant effect on processes of resonance fluorescence from atoms in the PC medium.

2. Resonance fluorescence from atom in PC's and vacuum modes

The recent experimental observations of the resonance fluorescence from self-assembled quantum dots, being atom-like two-level emitters, have shown [8] not only the asymmetry which for some detuning becomes very significant but also surprising sideband linewidths, while the standard theory predicts that the fluorescence spectrum should consist of a symmetric triplet. This discrepancy between the experiment and the theory could originate from the fact that the ordinary theory of the resonance fluorescence does not take into account the fact that the effect of the vacuum modes on the spectrum of resonance fluorescence may be significant.

In the standard theory it is assumed that in the case when ω_L is close to the frequency $\omega_R = \omega_e - \omega_g$ of the transition between two bound states $|g\rangle$ and $|e\rangle$ only the resonance mode is included into the strong atom-laser interaction. However, as it has been shown in Ref. [9], the vacuum modes that manifest themselves in nonradiative transitions between the dressed states can give a significant contribution to this interaction. In quantum optics the strong atom-laser interaction is usually described by using the two-level model and the rotating wave approximation (RWA) [10], and the dressed states are defined as eigenstates of the system Hamiltonian H_{RWA}: $|\pm, n\rangle = a_\pm |e, n\rangle + b_\pm |g, n\rangle$, $a_+ = b_- = \cos\theta_n$, $b_+ = -a_- = \sin\theta_n$, where $|e, n\rangle$ ($|g, n\rangle$) denotes the state containing the atom in the bare state $|e\rangle$ ($|g\rangle$) and n photons in the laser mode and θ_n is the mixing angle defined by $\tan(2\theta_n) = -\Omega_n/\Delta$ with Ω_n and Δ being the Rabi frequency $\Omega_n = 2g_L\sqrt{n+1}$ and detuning respectively. The energies of these states are $E_{\pm,n} = (n + 1/2)\omega_L + (\omega_e - \omega_g)/2 \pm \Omega_R^{(n)}/2$ with $\Omega_R^{(n)}$ being the generalized Rabi frequency. The states dressed by the interaction of the atom with the resonance laser field are not dressed by its interaction with the vacuum modes, and in our investigation of the problem we do not assume a priori that this interaction is weak in any case and we use nonperturbative methods of its description based on the the generalized dynamical equation (GDE), which [11] has been derived as a direct consequence of the first principles of quantum physics. Being equivalent to the Schrdinger equation in the case when the interaction in a quantum system is instantaneous, GDE allows one to extend quantum dynamics to the case of nonlocal-in-time interactions. This equation has been proven to be a useful tool for solving various problems in nuclear physics [12, 13], atomic physics [7, 14–17] and quantum optics [9, 18]. Within this approach the QED bound state problem is solved by using the formalism of the self-energy function $C_n(z)$ (see for details Ref. [19]), that describes the interaction of the particles in the state $|n\rangle$ with the vacuum. The self-energy QED corrections to the energies $E_n^{(0)}$ of the bare states are determined by the equation $z - E_n^{(0)} - C_n(z) = 0$ where $C_n(z)$ in turn are determined by GDE. In the first approximation the solution of this equation is $E_n \equiv E_n^{(0)} + C_n^{(0)}(E_n^{(0)}) \equiv E_n^{(0)} + \Delta E_n^L - \frac{i}{2}\Gamma_n$, where E_n is the energy of the real dressed state, E_n^L is the Lamb shift, Γ_n is the natural level width, and $C_n^{(0)}(z)$ is determined by the equation $dC_n^{(0)}(z)/dz = -\langle n|H_I \widetilde{G}_0^2(z) H_I|n\rangle$, with H_I being the QED Hamiltonian. For this approximation to be valid, probability amplitudes of the nonradiative transitions between the states with the same total angular momentum J, its projection J_z and parity must be much smaller than the energy distance between these states. In addition the variation of the function $C_n(z)$ with the energy must be negligible on the energy interval of order of the value of the Lamb shift. This takes place in the case when the atom is in free space. In fact, in this case the energy distance between the states with the same J, J_z and parity are of order of $\alpha^2 m_e$, while the probability amplitudes of nonradiative transitions between them are of order of $\alpha^5 m_e$, and the self-energy function varies significantly with the energy only on energy intervals of order m_e. The situation is dramatically changed in the case of an emitter subject to resonance laser field because the energy distance between dressed states $|+, n\rangle$ and $|-, n\rangle$ having the same J, J_z and parity equals to the Rabi frequency. As a result the nonradiative transitions between these states begin to play an important role. At leading order the probability amplitudes $\langle \pm, n|\Sigma(z)|\mp, n\rangle$ of these transitions are given by the formula [9] $\langle \pm, n|\Sigma(z)|\mp, n\rangle = \cos\theta \sin\theta (C_g^{(0)}(z) - C_e^{(0)}(z))$, where $C_g^{(0)}(z)$ and $C_e^{(0)}(z)$ are the self-energy functions of the bare states $|g\rangle$ and $|e\rangle$ respectively. At the same time, in the case of an atom being in free space the energy interval of order Ω_R is too small for the variation with energy of $\langle \pm, n|\Sigma(z)|\mp, n\rangle$ to be noticeable. In this case the energy dependence of $\langle \pm, n|\Sigma(z)|\mp, n\rangle$ can be neglected, and they can be regarded as matrix elements of an additional term in the Hamiltonian describing the interaction of the atom with resonance laser field. Adding such a term to the RWA Hamiltonian will lead only to a change in the mixing angle θ_n, and, as a consequence, will not give rise to asymmetry of the Mollow spectrum. However, as we show below, because of the singular behavior of the photon density of states the variation of self-energy functions $C_g(z)$ and $C_e(z)$ of an atom in the PC medium on an energy interval compared to the energy difference between the dressed states $|+,n\rangle$ and $|-,n\rangle$ can be very significant, and hence the vacuum modes are involved into the strong atom-laser interaction.

3. Self-energy functions of an electron in PC

Let us now investigate the behavior of the self-energy functions of an atom in the PC medium. Atomic self-energy functions are determined by the self-energy function of the atomic electron. The electron self-energy function can be determined in the same way that was used in [6] in determining the PC medium correction to the electron mass. In this way for the free electron with momentum $\mathbf{p}$ placed in the PC medium we get

$$C_{\mathbf{p}}^{pc}(z) = \sum_{\mathbf{p}'}\sum_{\mathbf{k},n} \frac{\langle \mathbf{p}|\, H_I^{pc}\, |\mathbf{p}';\mathbf{k},n\rangle \langle \mathbf{p}';\mathbf{k},n|\, H_I^{pc}\, |\mathbf{p}\rangle}{z - \frac{\mathbf{p}'^2}{2m_e} - \omega_{\mathbf{k}n}} \tag{1}$$

Assuming that $\mathbf{p} \gg \mathbf{k}+\mathbf{G}$ and replacing the discrete sums by integrals we get

$$C_{\mathbf{p}}^{pc}(z) = -\frac{\alpha}{2m_e^2\pi^2}\sum_n \int\limits_{FBZ} \frac{d^3k}{\omega_{\mathbf{k}n}} \frac{\sum_{\mathbf{G}} |\mathbf{p}\cdot\mathbf{E}_{\mathbf{k}n}(\mathbf{G})|^2}{z - \frac{\mathbf{p}^2}{2m_e} - \omega_{\mathbf{k}n}} \tag{2}$$

with $\mathbf{E}_{\mathbf{k}n}(\mathbf{G})$ being the coefficients in the plane-wave expansion $\mathbf{E}_{\mathbf{k}n}(\mathbf{r}) = \sum_{\mathbf{G}} \mathbf{E}_{\mathbf{k}n}(\mathbf{G})e^{i(\mathbf{k}+\mathbf{G})\cdot\mathbf{r}}$ of the Bloch eigenfunctions $\mathbf{E}_{\mathbf{k}n}(\mathbf{r})$. Here $\mathbf{G}$ is the reciprocal lattice vector of the photonic crystal ($\mathbf{G} = N_1\mathbf{b}_1 + N_2\mathbf{b}_2 + N_3\mathbf{b}_3$, where $\mathbf{b}_i$ are primitive basis vectors of a reciprocal lattice). Taking into account the polarization structure of the coefficients $\mathbf{E}_{\mathbf{k}n}(\mathbf{G}) = e_{\mathbf{k}n}^{\mathbf{G}1}\varepsilon_{\mathbf{G}1} + e_{\mathbf{k}n}^{\mathbf{G}2}\varepsilon_{\mathbf{G}2} + e_{\mathbf{k}n}^{\mathbf{G}3}\varepsilon_{\mathbf{G}3}$ we get

$$C_{\mathbf{p}}^{pc}(z) = -\frac{\alpha E_K}{m_e\pi^2}\sum_n \int\limits_{FBZ} \frac{d^3k}{\omega_{\mathbf{k}n}} \frac{\sum_{\mathbf{G},\lambda} \left|e_{\mathbf{k}n}^{\mathbf{G}\lambda}\right|^2 (\varepsilon_{\mathbf{G}\lambda})_p}{z - E_K - \omega_{\mathbf{k}n}}, \tag{3}$$

where $E_K = \frac{\mathbf{p}^2}{2m_e}$ and $(\varepsilon_{\mathbf{G}\lambda})_p$ are the projections of the polarization vectors onto the axis of the electron momentum. In order to eliminate these, we can assume that summing over $\mathbf{G}$ and λ averages these values, yielding a factor of $1/2$. Further estimates can be made with unknown solutions to Maxwell equations in a PC medium by building a model analogous to the one described in [6]. In this model Eq.(4) takes form

$$C_{\mathbf{p}}^{pc}(z) = -\frac{\alpha E_K}{m_e\pi^2}\int\limits_0^\Lambda \frac{d\omega}{\omega}\frac{N_p(\omega)}{z - E_K - \omega}, \tag{4}$$

where $N_p(\omega) = N_{DOS}(\omega)D(\omega)$, $N_{DOS}(\omega)$ is the photon density of states $N_{DOS}(\omega) = \frac{1}{4\pi}\sum\limits_n \int\limits_{FBZ} d^3k\delta(\omega - \omega_{\mathbf{k}n})$ and $D(\omega) = \frac{1}{2}\sum_{\mathbf{G},\lambda}\left|e_{\mathbf{k}n}^{\mathbf{G}\lambda}\right|^2_{|\omega_{\mathbf{k}n}=\omega}$. Here the cutoff Λ is imposed to separate the momentum region relevant for the variation of $C(z)$ at low energies. Real density of states and, as a consequence, $N_p(\omega)$ have very complicate behavior because of Van Hove singularities [20], see Fig 1(a). In the case when pole of the integrand in (4) is in the vicinity of this singularities, the value of the $C(z)$ can be very significant. To demonstrate this let us consider the influence of some critical points on the $C(z)$. For this we determine $N_p(\omega)$ so that it contains the same critical points as the real density of states.

As it follows from the results of calculations, the variation of the electron self-energy functions and as a consequence, the variation of the atomic self-energy functions $|g\rangle$ and $|e\rangle$ are very significant, and, as it follows from the comparisons of the results presented on Fig. 1(a) and Fig. 1(b) the singular behavior of the self-energy functions is a consequence of Van Hove singularities.

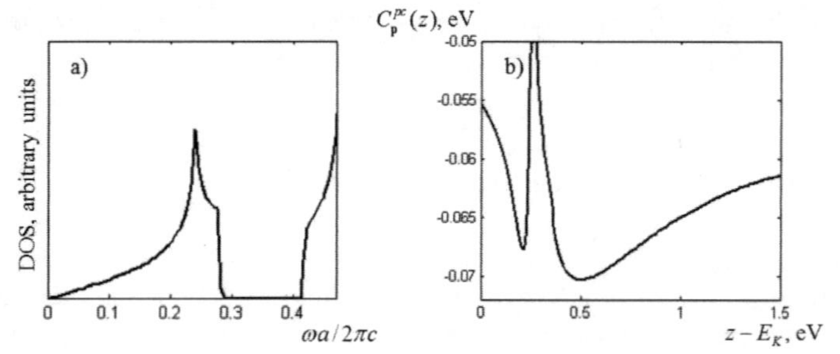

Figure 1. a) Density of states of the 2D PC composed of a hexagonal lattice of dielectric cylinders with permittivity $\epsilon = 12$ and radius $a/d = 0.2$ where d is period of PC. b) The self-energy function $C_{\mathbf{p}}^{pc}(z)$ of a free electron with $E_K = 3$ eV placed in PC medium calculated using Eq.(4)

4. Conclusion

We have shown that Van Hove singularities in the behavior of the photon density of states results in a singular behavior of the self-energy functions of the atomic states $|g\rangle$ and $|e\rangle$. This means that under certain conditions the vacuum modes are involved into the strong atom-laser interaction. Moreover, coming these modes into play can make this interaction much more stronger, and this can have a significant effect on processes of resonance fluorescence from atoms in the PC medium.

Acknowledgments

This work was funded by the subsidy allocated to Kazan Federal University for the state assignment in the sphere of scientific activities.

References

[1] Quang T, Woldeyohannes M, John S and Agarwal G S 1997 *Phys. Rev. Lett.* **79** 5238
[2] Bay S, Lambropoulos P, Mölmer K 1997 *Phys. Rev.* A **55** 1485
[3] Busch K, Vats N, John S, Sanders B C 2000 *Phys. Rev.* E **62** 4251
[4] Zhu S Y, Chen H, Huang H 1997 *Phys. Rev. Lett.* **79** 205
[5] John S and Quang T 1995 *Phys. Rev. Lett.* **74** 3419
[6] Gainutdinov R Kh, Khamadeev M A and Salakhov M Kh 2012 *Phys. Rev.* A **85** 053836
[7] Gainutdinov R Kh, Khamadeev M A and Salakhov M Kh 2013 *J. Phys.: Conf. Ser.* **478** 012017
[8] Vamivakas A N, Zhao Y, Lu C-Y and Atatüre M 2002 *Nat. Phys.* **5** 198
[9] Gainutdinov R Kh and Mutygullina A A 2009 *Physica Scripta* **T135** 014021
[10] Scully M and Zubairy M 1997 *Quantum Optics* (Cambridge: Cambridge University Press)
[11] Gainutdinov R Kh 1999 *J. Phys.* A **32** 5657
[12] Gainutdinov R Kh, Mutygullina A A, Khamadeev M A, Petrova A S 2013 *J. Phys.: Conf. Ser.* **478** 012019
[13] Gainutdinov R Kh, Moutygoullina A A 1997 *Physics of Atomic Nuclei* **60** 841
[14] Gainutdinov R Kh, Mutygullina A A, Scheid W 2001 *Proc. of SPIE* **4605** 214
[15] Gainutdinov R Kh, Mutygullina A A, Scheid W 2001 *Proc. of SPIE* **4605** 222
[16] Gainutdinov R Kh 1995 *ZETP* **108** 1600
[17] Gainutdinov R Kh 1986 *Optika i Spektroskopiya* **60** 890
[18] Gainutdinov R Kh, Mutygullina A A 2013 *J. Phys.: Conf. Ser.* **478** 012018
[19] Gainutdinov R Kh 1984 *Ukrainskii fizicheskii zhurnal* **29** 805
[20] Fussell D P, McPhedran R C, Martijn de Sterke C 2005 *Phys. Rev.* E **72** 046605

Laser field distribution near inclined taper optical antenna

A R Gazizov, A M Rogov, S S Kharintsev and M Kh Salakhov

Kazan Federal University, 18, Kremlevskaja str., 420008 Kazan, Russian Federation

E-mail: Almaz.Gazizov@kpfu.ru

Abstract. This work focuses on the mechanism of laser field enhancement within a gap between a taper plasmonic nanoantenna and a dielectric substrate in the inverted optical configuration. We show that field enhancement factor depends on several parameters such as: tip-sample distance, radius of curvature of the tip and its inclination angle in respect to the substrate. In particular, a relation between the field enhancement near the tip apex and the inclination angle is deduced. Numerical simulation and optimization were performed within the framework of the FDTD method and the Green function formalism.

1. Introduction

In the recent years, there has been great interest in optical antennas design [1]. Optical antenna is a device that converts freely propagating radiation into localized modes and vice versa [2]. Optical antennas allow one to probe a sample at a subwavelength scale via different optical near-field microscopy techniques [3]. One of the methods is, for example, tip-enhanced Raman scattering (TERS), which provides a non-destructive chemical analysis of subwavelength structures in the visible range under normal conditions. In this method the nanoantenna is positioned by a distance of several nanometers above the sample surface in highly focused laser field, providing amplification of interaction between optical radiation and matter, as well as strong localization of electromagnetic field [4].

One of the goals of the optical near-field microscopy is an increase of field enhancement and concentration via optical antenna design. Recent results have been obtained using different physical phenomena that help to reach the highest electric field enhancement, for example, scanning probe microscopy [6,7,9] or scanning tunneling microscopy [5,8]. The optical resolution of ~2 nm was obtained in the experiment with the use of ultra-low temperature and ultra-high vacuum [5]. In the Ref. [6,7] optical resolution of several nanometers was obtained under normal conditions, but with the use of additional processes.

Sub-10 nm optical resolution may be achieved by additional effects. In Ref.[10,11], for example, there were performed experiment and simulation of the excitation of surface plasmon polaritons on a metal grid side surface of the probe and the adiabatic compression near the tip.

In this paper we describe physical effects that occur in tip enhanced near-field optical microscopy when inclined tip is used. This method is practically feasible and meaningful; via this method we can increase the gain of the optical response of the sample and increase optical resolution, and make steps towards the development of new devices using the phenomenon of surface and localized plasmon resonances in optical near-field interaction for processing and transmitting information. We show that inclined tips application provides high concentration of energy and enhancement of intensity larger than in use of vertical probe. The relation between the field intensity enhancement and the angle of inclination in XZ-plane by using finite-difference time-domain simulation is analyzed in detail. Our aim is to study physical effects that occur in the process of scattering of highly focused laser light with the inclined tip that is positioned by a distance of several nanometers above the substrate surface.

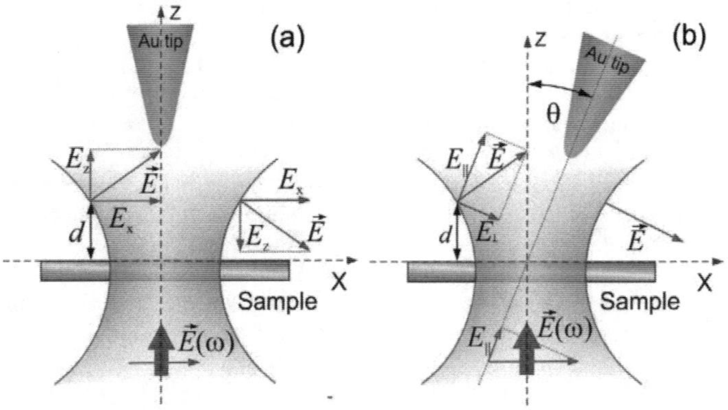

Figure 1. Tip orientation relatively to the optical z-axis: (a) 0°;(b)at θ angle [12].

2. Results and discussion

Figure 1(b) shows the schematics of the problem. To find the relation between the electric field intensity near tip apex and the angle of inclination of tip we use the Green's function formalism described in Ref [13-15]. A total electric field scattered with a metallic nanosphere with a radius ρ, which is placed at the distance Δ from the planar interface between two dielectric media with dielectric constants ε_1 and ε_2, reads as

$$\mathbf{E}(\mathbf{r},\omega) = \mathbf{E}_0(\mathbf{r},\omega) + \frac{\omega^2}{\varepsilon_0 c^2}\overleftrightarrow{\mathbf{G}}_0(\mathbf{r},\mathbf{r_0};\omega)\alpha_{\text{eff}}\mathbf{E_0}(\mathbf{r_0},\omega), \quad (1)$$

here $\mathbf{E}_0$ is incident laser field, ω is an angular frequency of the incident radiation, $\overleftrightarrow{\mathbf{G}}_0(\mathbf{r},\mathbf{r_0};\omega)$ is the Green's function in free space and α_{eff} is effective polarizability of inclined tip which determines the influence of the image field from the dielectric interface as in the expression [16]:

$$\mathbf{p}_{\text{eff}} = \alpha[\mathbf{E}_0 + \mathbf{E}_{\text{image}}] \equiv \alpha_{\text{eff}}\mathbf{E}_0, \quad (2)$$

here $\mathbf{E}_{\text{image}}$ is a local field of the image dipole, which can be written as

$$\mathbf{E}_{\text{image}} = \frac{1}{4\pi\varepsilon_0}\left(\frac{3(\mathbf{p}_{\text{eff}}(\mathbf{r}_0' - \mathbf{r}_0))(\mathbf{r}_0' - \mathbf{r}_0)}{|\mathbf{r}_0' - \mathbf{r}_0|^5} - \frac{\mathbf{p}_{\text{eff}}}{|\mathbf{r}_0' - \mathbf{r}_0|^3}\right), \quad (3)$$

where $\mathbf{r}_0'$ is a vector position of an image dipole.

The polarizability of the small metallic sphere α in Eq. (2) could be written as [17]

$$\alpha = 4\pi\varepsilon_0\rho^3 \begin{pmatrix} \frac{\varepsilon_m - 1}{\varepsilon_m + 2} & 0 & 0 \\ 0 & \frac{\varepsilon_m - 1}{\varepsilon_m + 2} & 0 \\ 0 & 0 & f_e/2 \end{pmatrix}, \tag{4}$$

here f_e is a complex field enhancement factor depending on tip material and geometry, ε_m is a dielectric function of the sphere. Since α_{xx} and α_{yy} components of the polarizability tensor much less than α_{zz} component, in our approach we make the approximation $\alpha_{xx} = \alpha_{yy} = 0$.

Then longitudinal effective polarizability of inclined tip reads as

$$\alpha_{\text{eff}}^{\|} = \frac{2\pi\varepsilon_0\rho^3 f_e \cos\theta'}{1 + \rho^3 b \dfrac{f_e(1 - 3\cos^2\theta)}{16(\Delta + \rho)^3}}, \tag{5}$$

where $b = \frac{\varepsilon_2 - \varepsilon_1}{\varepsilon_2 + \varepsilon_1}$ is a quasi-static Fresnel-reflection coefficient, θ is an angle of inclination of the probe, θ' is an angle between the electric field and the probe symmetry axis. Analytic angular dependence of the electric field appears due to the presence of the non-diagonal components in polarizability tensor of the optical antenna rotated in the space which depend on the angle of inclination. There is an enhancement of the only field projection on the probe symmetry axis (approximately true), so the polarizability tensor has only one component of a nonzero in its own coordinate system. Changes in the local fields caused by the substrate are equivalent to the modification of tip polarizability.

One of the experimental schemes in near-field imaging is the inverted optical configuration, in which the sample is illuminated with laser beam propagating normally to the sample surface (Z-axis) is strongly focused near planar surface of the sample. Strong focusing of the beam is provided by use of immersion objective lenses with $NA > 1$ [18]. Herewith there are the regions in the focal where the field is polarized longitudinally (along the beam propagation direction), referred to as lobes [19]. Their quantity, their pattern and their field intensity are dependent on the numerical aperture, the polarization and the mode of the incident laser beam. If the beam is linearly polarized there are two regions in the focal plane located along the polarization axis [18].

Since the antenna mainly interacts with longitudinal lobes near vicinity of the tip we may drop X- and Y-components for the scattered field in tip's own coordinate system. In this case we substitute $|\mathbf{r} - \mathbf{r_0}| = \Delta + \rho$. Neglecting the contribution of the external field $\mathbf{E}_0$ compared to the enhanced field, we find the field in the vicinity of the tip apex:

$$E(\mathbf{r}_0, \omega) = \frac{f_e \cos\theta'}{1 + \rho^3 b \dfrac{f_e(1 - 3\cos^2\theta)}{16(\Delta + \rho)^3}} E_0(\mathbf{r_0}, \omega) e^{-ik\rho}. \tag{6}$$

Since the value of the external field (the magnitude of the vector) the same in the right and left lobes, the ratio of the intensities at the maximal power:

$$\left|\frac{E_{L\max}}{E_{R\max}}\right|^2 = \frac{\cos^2\theta_1'}{\cos^2\theta_2'} = \frac{\cos^2(\eta_1 - \theta)}{\cos^2(\pi - \eta_1 - \theta)}, \tag{7}$$

Journal of Physics: Conference Series **560** (2014) 012008 doi:10.1088/1742-6596/560/1/012008

where η_1 is the parameter corresponding to the inclination angle of the electric field vector to the normal of the focal plane in the points of intensity maxima of the lobes.

Series of FDTD simulations were carried out for different inclination angles and positions of gold probe – scanning of the focal field along the axis of field polarization (X axis) with the increment of 10 nm. For this purpose we used the following tip parameters: tip apex curvature radius $\rho = 13\,\text{nm}$, the apex angle $\alpha = 30°$, tip-substrate distance $\Delta = 3\,\text{nm}$. Frequency of the incident light was $f = 450\,\text{THz}$ (in the resonance band of gold), the numerical aperture $NA = 1{,}4$. For numerical simulation the dielectric function of gold was taken from Ref. [20].

Figure 2 shows the results of parametric FDTD modeling of the ratio of maximum intensities in lobes on the probe angle.

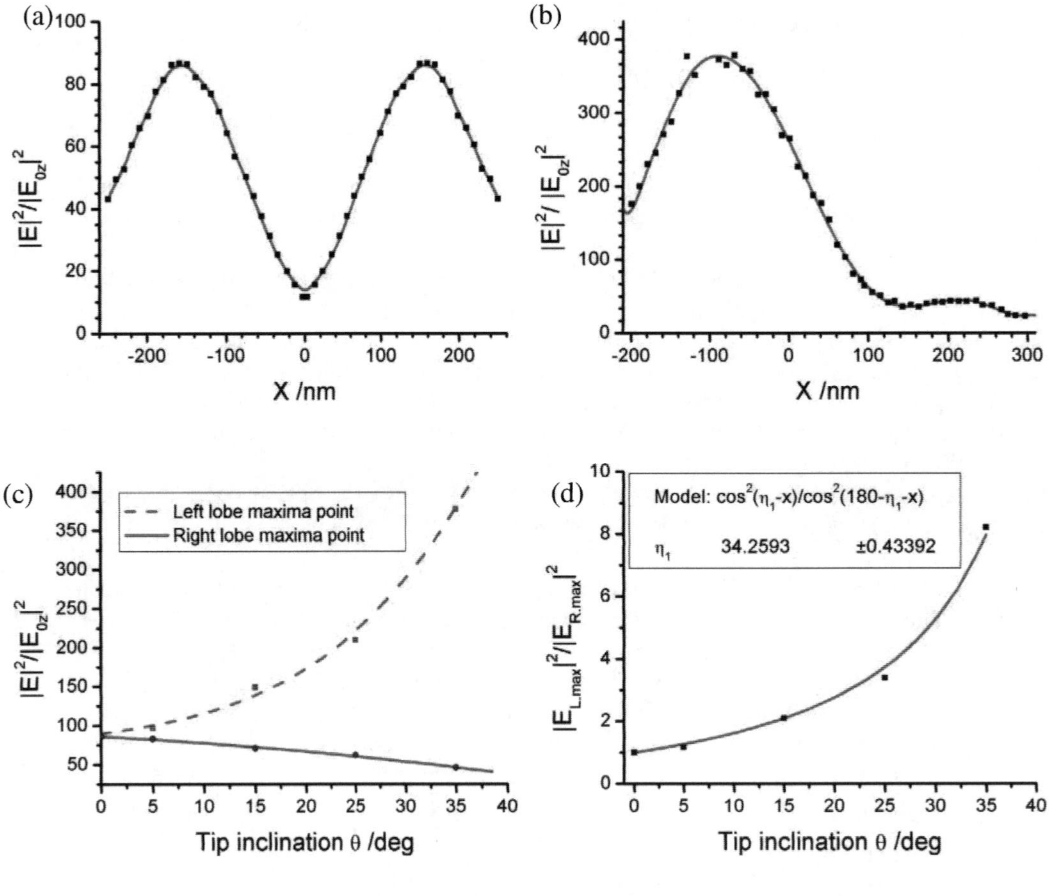

Figure 2. Field enhancement vs tip position when tip inclination of 0° (a) and 35° (b). Maximal field enhancement vs tip inclination (c). The ratio of the intensities in two lobes (left to right) vs inclination angle (d) (solid red line is an approximation with Eq. (7)).

For the purposes of our approximation we state *a priori* the final relation between the maximum intensity in two lobes and the angle of the probe in the form of Eq.(7). It is shown that the lobes are shifting to the right while the right lobe decreases in intensity and becomes difficult to distinguish at angles greater than 40°. Approximation gives the estimated value of the parameter $\eta_1 = 34°$. This means that at a this angle of inclination of the axis of the probe tip is oriented along the field vector direction and at the angle of $90-34=56°$ the right lobe disappears, and amplification of transverse component of the field occurs at large angles predominantly. Thus, the critical angle at which one of the lobes disappears with given parameters of the problem ($NA = 1{,}4$, $f = 450\,\text{THz}$) is obtained roughly equal to 56°.

3. Conclusion

We have studied effects of the field enhancement beneath metallic tapered optical antenna that is inclined in respect to the substrate. We have simulated an optical response of the inclined optical antenna and the relation between the field enhancement near the tip apex and inclination angle was found. In particular we have shown that the laser field lobe shifts towards the optical axis and becomes spread. Thus, asymmetric energy distribution behavior is observed. This phenomenon explains why even upright probe retained small asymmetry of enhancement in the lobes, because ideally upright clutch of the optical antenna to a tuning fork is impossible.

Acknowledgements

This work was funded by the subsidy allocated to Kazan Federal University for the state assignment in the sphere of scientific activities. This work was financially supported by the Russian Foundation for Basic Research (No. 13-02-00758 A). This work was done using equipment of Federal Center of Shared Equipment of Kazan Federal University.

References

[1] Novotny L 2011 *Physics today* **64** 47-52
[2] Novotny L and van Hulst N 2011*Nature Photonics* **5** 83-90
[3] Kharintsev S S, Hoffmann G G, Dorozhkin P S, de With G and Loos J 2007 *Nanotechnology* **18** 315502
[4] Georgi C and. Hartschuh A 2010 *Appl. Phys. Lett.* **97** 143117
[5] Chen C, Hayazawa N, Kawata S 2014 *Nature Communications* **5** 3312
[6] Yano T, Verma P, Saito Y, Ichimura T and Kawata S 2009 *Nature Photonics* **3** 473-477
[7] De Angelis F, Das G, Candeloro P *et al* 2009 *Nature Nanotechnology* **5** 67-72
[8] Stadler J, Schmid T and Zenobi R 2011 *ACS Nano* **5** 8442-8448
[9] Chan K L A and Kazarian S G 2010 *Nanotechnology* **21** 445704
[10] Berweger S, Atkin J M, Olmon R L and Raschke M B 2010 *J. Phys. Chem. Lett.* **1** 3427-3432
[11] Giugni A, Torre B Toma A *et al* 2013 *Nature Nanotechnology* **8** 845-852
[12] Kharintsev S S, Rogov A M and Kazarian S G 2013 *Rev. Sci. Instr.* **84** 093106
[13] Cançado L G, Jorio A, Ismach A, Joselevich E, Hartschuh A and Novotny L 2009 *Phys. Rev.Lett.* **103** 186101
[14] Maximiano R V, Beams R, Novotny L, Jorio A and Cançado L G 2012 *Phys. Rev. B* **85** 235434
[15] Rogov A M, Gazizov A R, Kharintsev S S, Salakhov M Kh 2013 *J. Phys.: Conf. Ser.* **478** 012015
[16] Novotny L and Stranick S J 2006 *Annu. Rev. Phys. Chem.* **57** 303-331
[17] Bouhelier A, Beversluis M, Hartschuh A and Novotny L 2003 *Phys. Rev. Lett.* **90** 013903
[18] Novotny L and Hecht B 2006 *Cambridge University Press* 559
[19] Bouhelier A, Beversluis M R, Novotny L 2003 *Appl. Phys. Lett.* **82** 4596-4598
[20] Johnson P B and Christy RW 1972 *Phys. Rev. B* **6** 4370-4379

Journal of Physics: Conference Series **560** (2014) 012009 doi:10.1088/1742-6596/560/1/012009

New approach of decomposition of complex spectral contours based on particles swarm optimization

D Z Galimullin, M E Sibgatullin[1], D I Kamalova, M Kh Salakhov

Kazan Federal University, 420008, Kremlevskaja str., 18, Kazan, Russian Federation

E-mail: sibmans@mail.ru

Abstract. The approach based on the stochastic algorithm of particle swarm optimization was used for mathematical treatment of spectral contours. It was shown that the method supposed allows one to separate complex model spectra and to determine parameters of spectral components. In the mathematical experiments a random fractal noise as a model of noise was applied.

1. Introduction

The problem of decomposition of complex spectral contours into components relates to problems of optimization which consist in evaluation of the entrance parameters when the objective function reaches the extremum. We suggest a new approach to the problem of decomposition of complex optical spectra which is based on the particle swarm optimization (PSO) algorithm. This method simulates particles population behavior in the space of optimization parameters [1]. The knowledge of exact gradient of function under optimization is not required. In this work the method supposed is shown to allow one to separate complex model spectra and to determine parameters of components.

2. Particle swarm optimization algorithm

The method simulates a multi-agent system in which agents-particles move to optimal solutions by exchanging information with their neighbors [1]. The current position of a particle is characterized by coordinates $\vec{x}$ in a set of solutions and by velocity vector $\vec{v}$. These parameters are chosen as random at the stage of initialization. At the next step each particle keeps both the coordinates of the best solution $\vec{p}$ found earlier and the best solution $\vec{g}$ found for all particles. When the algorithm is acting, the direction and length of the velocity vector of each particle changes in accordance with data of obtained optimum [2]

$$\vec{v} = w\vec{v} + a_1 r_p (\vec{p} - \vec{x}) + a_2 r_g (\vec{g} - \vec{x}), \quad (1)$$

where $\vec{v}$ is the vector of the particle velocity, w is the inertia coefficient which characterizes the balance between the search area and the computed suboptimal solutions, a_1 and a_2 are acceleration coefficients, $\vec{p}$ is the best position found by a particle, $\vec{g}$ is the best position found by all the particles of the system, $\vec{x}$ is the vector of particle coordinates, r_p and r_g are random numbers between 0 and 1. When the direction of $\vec{v}$ is calculated, the particle moves to the point

[1] To whom any correspondence should be addressed.

$$\vec{x} = \vec{x} + \vec{v}t, \quad (2)$$

It is possible to accept value t equal to 1 since time is a conventional constant value which does not influence on the algorithm. The values of the best points are then renewed both for each particle and for all particles in ensemble. The cycle is repeated after that until the stopping criterion is executed.
The first step of PSO is to create a particle swarm in a multi-dimensional search space where each component characterizes one of the desired parameters of the components of a complex spectrum. Accordingly, for a spectral contour of four components (each component is described by three parameters), the search space is 12-dimensional. Then, the initial swarm position is defined as random. It is the parametric residual function between the initial data and a randomly specified. The initial values of velocity vectors are equal to zero. The steps of the procedure are calculation of the velocity vectors of all particles using Eq. (1), renewal of their coordinates with Eq. (2), calculation of the best solution for each particle at the current step and the best solution of whole swarm at the current iteration. We give a new coefficient of inertia by reducing it with a certain constant and thus minimizing the displacement of the particles when increasing a number of iterations. This guarantees rapid convergence of the algorithm. Then, the cycle is repeated until the exit criterion is executed: the residual minimum of the function is achieved or the certain number of iterations is performed.

3. Results and discussion

To investigate the possibility of recovering the complex contours by PSO algorithm, we have utilized a simulated spectrum consisting of four components of Gaussian form. This model is useful for description of real physical processes. Figure 1 shows the simulated profile composed of four Gaussians and reconstructed by means of PSO. The initial (integral) and reconstructed profiles coincide with the error not higher than 1–2% during each execution of the algorithm.

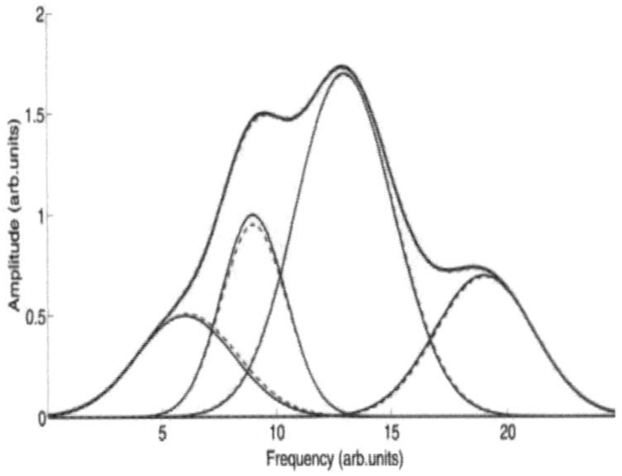

Figure 1. Reconstruction of a model profile by means of PSO: dotted line – model Gaussians and integral spectrum; solid line – reconstructed components and integral spectrum

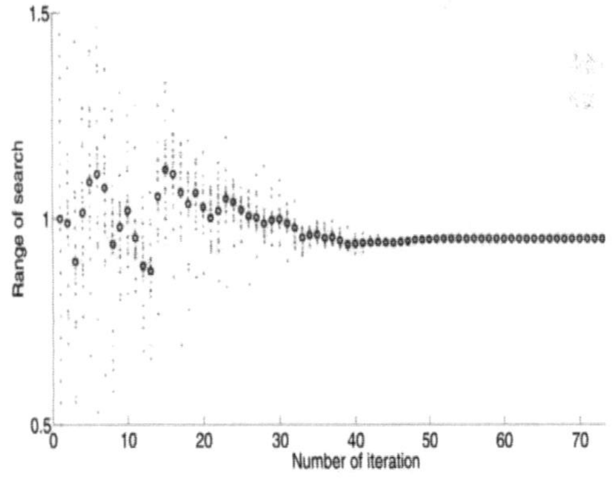

Figure 2. Dynamics of the average value (circles) and coordinates (dots) of a particle swarm when calculating the amplitude versus the iteration number

Figure 2 shows the amplitude of one of the components (with the amplitude normalized to 1) as a function of the number of iterations. The number of particles in the swarm is equal to 16 (small dots). As it can be seen, the average coordinate of the particle swarm that describes the amplitude of the Gaussian virtually did not deviate from the preset initial unit value beginning with the 40th iteration. Similar dependences were derived for all target parameters of the spectral profile components.

Note that the algorithm does not require fixed initial conditions, since it is sufficient to specify the search space once the particles are homogeneously distributed there.

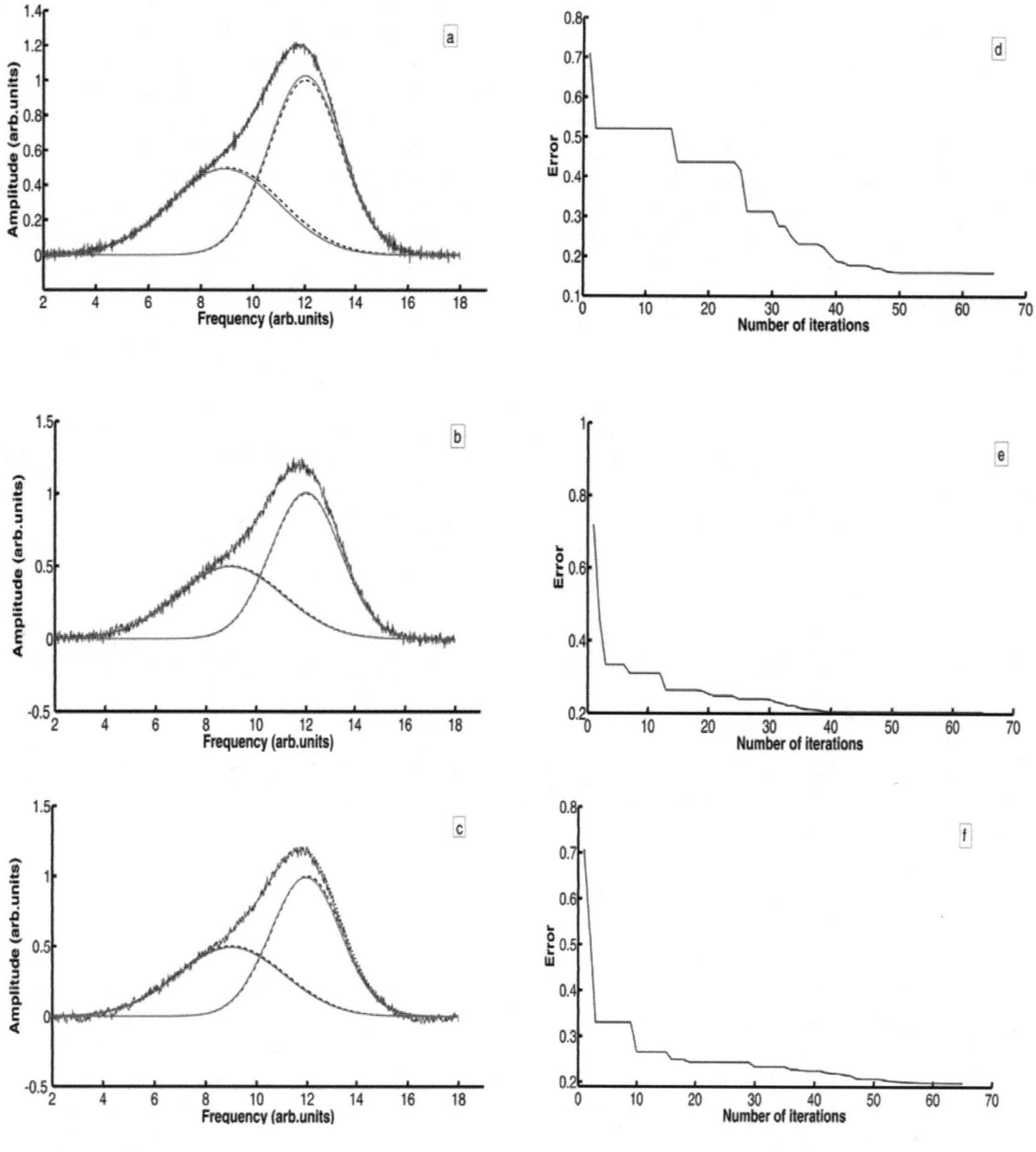

Figure 3. Reconstruction of a simulated profile having two Gaussians and the relative error as a function of the iteration number for different Hurst index: H=0.1 (*a, d*), H=0.5 (*b, e*), H=0.9 (*c, f*). The noise level is equal to 5%.

Analyzing experimental data researchers face the problem when noise corrupts the signal. We have studied the stability of PSO algorithm with respect to noise and also we have investigated influence of the character of noise on the decomposition of spectral contours. In our mathematical experiments, we used a random fractal noise as the model of noise [3]

$$\xi_f = D^{-(H-1/2)}\xi_g,$$

where ξ_g is a non-correlated Gauss random process, D^{-v} is the fractional operator of the order v, H is the Hurst index. Random fractal processes occupy an intermediate position between deterministic and accidental [4, 5]. This model describes well curves of real physical processes and is useful in analyzing experimental noise. Varying the magnitudes of Hurst index allow to simulate a set of noises with different frequency structure from high-frequency (H=0.1) to low-frequency structure (H=0.9).
Figure 3 shows the results of recovery of simulated contour of two Gaussian components with a noise level of 5% and different Hurst index: H=0.1 and 0.9 ("color" noise), H=0.5 ("white" noise). Besides, Figure 3 shows the relative error as the difference between the areas of the initial and reconstructed profiles with respect to the area of the initial profile as a function of the iteration number. The number of iterations was equal to 65 when influence of noise has been analyzed.
The use of PSO algorithm for the reconstruction of complex contours with noise of various structures shows that the results are in good agreement with the initial simulation components. By comparison of obtained results one may conclude that the deconvolution of complex spectral signal is the best at H=0.5. So, the best solution of deconvolution takes place when a spectral contour is distorted by statistically independent noise. Presence of correlated noise in a signal worsens reconstruction slightly. However, the results still remain at high level of accuracy.
Thus, the approach based on the stochastic algorithm of PSO for mathematical treatment of spectral contours allows one to separate complex spectra with noises of different structure and to determine parameters of spectral components.

Acknowledgements
This work was funded by the subsidy allocated to Kazan Federal University for the state assignment in the sphere of scientific activities.

References

[1] Kennedy J and Eberhart R 1995 *Proc. IEEE Int. Conf. on Neural Networks* **4** 1942
[2] Shi Y and Eberhart R 1998 *Proc. IEEE Int. Conf. on Evolutionary Computation* 69
[3] Kharintsev S S *et al* 2000 *JQSRT* **67** 239
[4] Mandelbrot B 1982 *Fractal Geometry of Nature*
[5] Pallikari F 2001 *Chaos. Solutions Fractals* **12** 1499

Ce^{3+}, Pr^{3+}:$LiY_{0.3}Lu_{0.7}F_4$ Mixed Crystal as a Perspective Upconversionally Pumped UV Active Medium

V G Gorieva, V V Semashko, S L Korableva, M A Marisov, V V Pavlov

Kazan Federal University, 420008, Kremlevskaja str., 18, Kazan, Russian Federation

E-mail: ekorre_v@mail.ru

Abstract. The key parameters of processes provided up-conversion pumping of 5d-states of Ce^{3+} ions in Ce^{3+}, Pr^{3+}:$LiY_{0.3}Lu_{0.7}F_4$ mixed crystals were estimated for the first time. The prospects of this material as the first up-convertionally pumped UV active media were demonstrated.

1. Introduction

The most effective solid-state UV lasers are based on Ce:$LiCaAlF_6$ and Ce:$LiLuF_4$ active media and operated on interconfigurational 5d-4f transitions of Ce^{3+} ions [1]. However, UV pumping radiation of these lasers induces degradation of optical properties of solid-state media because of various photodynamic processes (PDP). One of the ways to avoid or significantly reduce harmful manifestations of PDP is to use up-conversion pumping [2].

Here we investigate an opportunity of effective pumping of 5d-states of Ce^{3+} ions in $LiY_{0.3}Lu_{0.7}F_4$ (LYLF) crystals by stepwise $^3H_4 \rightarrow {}^1D_2, {}^3Pj \rightarrow 4f5d$ up-conversion excitation of states of 4f5d-configuration of Pr^{3+} ions, followed by the energy transfer from Pr^{3+} to Ce^{3+} ions.

2. Results and discussion

The choice motivation of Ce^{3+},Pr^{3+}:$LiY_{0.3}Lu_{0.7}F_4$: mixture crystals follows from the results of ref.[3], which show the necessity to increase the concentrations of the impurity ions for effective excitation energy transfer from Pr^{3+} to Ce^{3+} ions. Take into account that the scheelite crystal structure persists throughout the series of crystals with the formula $LiY_xLu_{1-x}F_4$, where x is any value from 0 to 1 and the $LiY_{0.3}Lu_{0.7}F_4$ (LYLF) crystals have the greatest isomorphic capacity to Pr^{3+} and Ce^{3+} ions [4], samples of Ce,Pr:LYLF mixture crystals were grown in Kazan University. The Pr^{3+} ions content in the initial components was 1 at% and the concentration of Ce^{3+} ions was varied from sample to sample from 0 to 2 at%. The real concentrations of the dopants in the samples (see Table 1) were determined by X-ray fluorescence and optical absorption spectroscopic methods described elsewhere [5].

Table 1.Absolute concentration of Pr^{3+} and Ce^{3+} ions, fluorescent lifetimes of the lowest 4f5d-state of Pr^{3+} ions and the coefficients of excitation energy transfer from Pr^{3+} ions to Ce^{3+} ions in the $LiY_{0.3}Lu_{0.7}F_4$ crystal samples

№ of the sample	Concentration of Pr^{3+} ions (C_{Pr}), at%	Concentration of Ce^{3+} ions (C_{Ce}), at %	$\tau_{люм}$, ns	K*10^7,c^{-1}
1	0.51 ± 0.05	0	20±1,1	-
2	0.38 ± 0.07	0.16 ± 0.04	14,6±0,8	1.85
3	0.38 ± 0.05	0.28 ± 0.05	13,9±0,9	2.19
4	0.24 ± 0.04	0.47 ± 0.06	16,0±0,8	1.25

Typical absorption and luminescence spectra of Pr^{3+} and Ce^{3+} ions in the crystal samples are shown in Figure 1(a) and typical kinetics of 4f5d-4f^2 luminescence of Pr^{3+} in LYLF crystals, as well as the temporal profile of the laser excitation pulse are shown in Figure 1(b). Determination of excitation energy transfer efficiency from Pr^{3+} to Ce^{3+} ions was carried out by comparing the characteristic decay time of the luminescence of the donor (Pr^{3+}) depending on the concentration of acceptor ions (Ce^{3+}) [6]. The excitation of 4f5d-4f^2 luminescence of Pr^{3+} ions in LYLF crystals were performed at 218 nm. Fluorescence decays of Pr^{3+} ions were studied at 255 nm using the photomultiplier PMT-87 and digital oscilloscope DPO 7354 Tektronix.

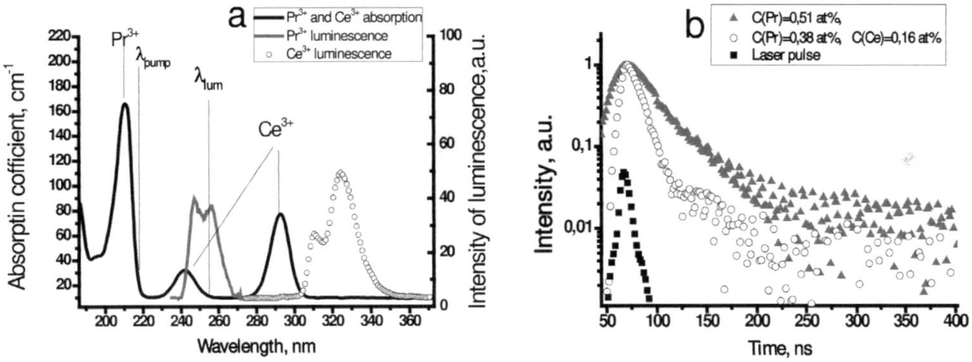

Figure 1. Absorption and luminescence spectra of Pr^{3+} and Ce^{3+} ions in $LiY_{0.3}Lu_{0.7}F_4$ crystals (a) and typical kinetics of 4f5d-4f^2 luminescence of Pr^{3+} ions at 255 nm in crystals with temporal profile of the laser excitation pulse (b)

Energy transfer parameter k from 4f5d-states of Pr^{3+} ions to 5d-states of Ce^{3+} ions were estimated by formula (1):

$$K = 1/\tau_{codoped} - 1/\tau_0 , \quad (1)$$

where τ_0 –decay time of 4f5d-4f^2 luminescence of Pr^{3+} in $LiY_{0.3}Lu_{0.7}F_4$ crystals without doping by Ce^{3+} ions, $\tau_{codoped}$ – with the one, correspondently. The results are shown in Table 1. As it can be seen from the table, co-doping of the Pr:LYLF crystals by Ce^{3+} ions leads to displacement of Pr^{3+} ions and that is why the maximum of energy transfer coefficient reaches 2.19*10^7 c^{-1} in the sample #3 with total content of Pr^{3+} and Ce^{3+} ions 0.38 and 0.28 at.%, correspondently. This result exceeds the value of the energy transfer coefficient previously defined in [3] for Ce^{3+}, Pr^{3+}:$LiLuF_4$ crystals.

To excite the states of 5d-configuration of Ce^{3+} ions the double-stepwise up-conversion scheme using intermediate $^1D_2$ manifold of Pr^{3+} ions was chosen. This scheme proposes to excite samples by two pumping radiation beams with 595 nm and 266 nm wavelengths [3]. The reason was the fact that, in accordance with data of [7], the color centers that are induced in Ce:YLF crystal by UV pumping (YLF is the "parent crystal" for the ones studied here) are discoloring by 266 nm irradiation of the samples. The experimental setup, the up-conversion scheme and legends are shown in Fig 2.

To estimate the perspectives of Ce,Pr:LYLF crystals as up-convertionally pumped UV active media, the dependences of integral intensity of Ce^{3+} luminescence on energy density at 266 nm for fixed energy density at 595 nm were studied. The model is depicted in Figure 2.

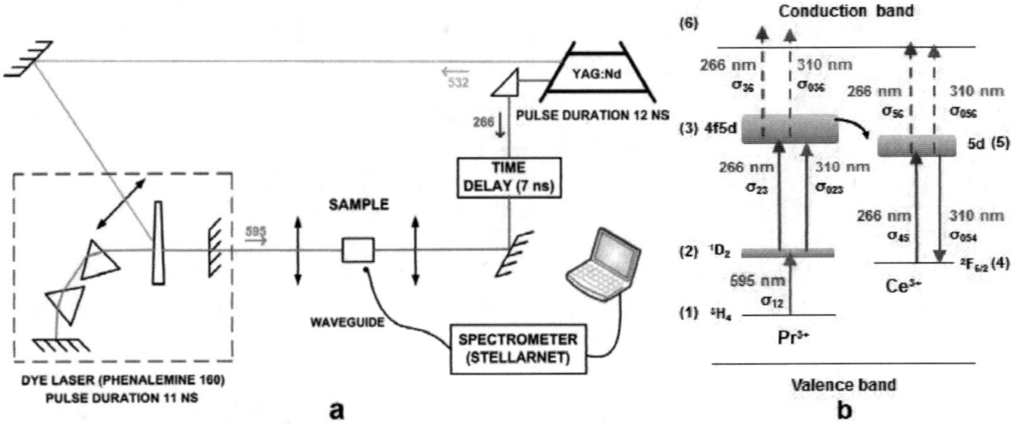

Figure 2. Experimental setup (a) and model of excitation of 5d-4f luminescence of Ce^{3+} ions in $LiY_{0.3}Lu_{0.7}F_4$ crystals (b). The solid arrows pointing upwards shows the transitions that provide excitation of 4f5d-configuration of Pr^{3+} ion for the two stepwise up-conversion pumping schemes and 5d-states of Ce^{3+} ions. Solid arrows pointing downwards show 5d-4f transitions of Ce^{3+} ion. Dashed arrows indicate the possible transitions from the 4f5d-excited states of Pr^{3+} ions and 5d-states of Ce^{3+} ions to the conduction band, arc arrow indicates energy transfer processes from Pr^{3+} to Ce^{3+} ions

Figure 2(b) was applied to interpret these dependences. Here the levels 1,4 are ground states of Pr^{3+} and Ce^{3+} ions; 2, 3, 5 – excited levels of Pr^{3+} and Ce^{3+} ions; 6 – conduction band of the crystal, correspondently. The σ_{12}, σ_{23} are absorption cross-sections from ground state to $^1D_2$ level and from $^1D_2$ to 4f5d level of Pr^{3+} ion at 266 nm, respectively; σ_{45} – absorption cross-section from the ground to 5d level of Ce^{3+} ion at 266 nm; σ_{36}, σ_{56}, σ_{036}, σ_{056} - absorption cross-sections from excited 4f5d and 5d levels of Pr^{3+} and Ce^{3+} ions at 266 nm; σ_{056} - absorption cross-sections from excited 5d level of Ce^{3+} ion at 310 nm; σ_{054} – stimulated 4f-5d transitions cross-sections of Ce^{3+} ions. These designations are also used in the equations (2) that describe this model:

$$\begin{aligned}
\frac{dn_1}{dtC} &= -U_1(t)\cdot\sigma_{12}\cdot n_1 + \frac{1}{\tau_2}\cdot n_2 + \frac{1}{\tau_3}\cdot n_3 + K\cdot n_3\cdot\frac{n_4}{Ce},\\
\frac{dn_2}{dt} &= U_1(t)\cdot\sigma_{12}\cdot n_1 - \frac{1}{\tau_2}\cdot n_2 - U_2(t)\cdot\sigma_{23}\cdot n_2 + U_2(t)\cdot\sigma_{32}\cdot n_3,\\
\frac{dn_3}{dtC} &= U_2(t)\cdot\sigma_{23}\cdot n_2 - U_2(t)\cdot\sigma_{32}\cdot n_3 - U_2(t)\cdot\sigma_{36}\cdot n_3 - \frac{1}{\tau_3}\cdot n_3 - K\cdot n_3\frac{n_4}{Ce} + p_{63}\cdot n_6,\\
\frac{dn_4}{dtC} &= -K\cdot n_3\frac{n_4}{Ce} - U_2(t)\cdot\sigma_{45}\cdot n_4 + \frac{1}{\tau_5}\cdot n_5,\\
\frac{dn_5}{dtC} &= K\cdot n_3\frac{n_4}{Ce} + U_2(t)\cdot\sigma_{45}\cdot n_4 - \frac{1}{\tau_5}\cdot n_5 - U_2(t)\cdot\sigma_{56}\cdot n_5 + p_{65}\cdot n_6,\\
\frac{dn_6}{dt} &= U_2(t)\cdot\sigma_{36}\cdot n_3 + U_2(t)\cdot\sigma_{56}\cdot n_5 - (p_{63}+p_{65})\cdot n_6,
\end{aligned} \tag{2}$$

where n_1, n_2, n_3, n_4, n_5 – populations of the appropriated states, $U_1(t)$, $U_2(t)$ – photon flux density of the pumping radiation at 595 and 266 nm, respectively; τ_2,τ_3,τ_5 – lifetimes of $^1D_2$, 4f5d states of Pr^{3+} ions and 5d states of Ce^{3+} ions, respectively; σ_{ij} – appropriated transition cross-sections, p_{63} and p_{65} – recombination probabilities, K – energy transfer coefficient from Pr^{3+} to Ce^{3+} ions; C_{Ce} and C_{Pr} – concentration of Ce^{3+} and Pr^{3+} ions.

Parameters τ_2, σ_{12}, σ_{23} are known from [8], τ_3,τ_5 – [5], [9], σ_{56} – from [10] and σ_{036}, p_{63} and p_{65} are estimated in [1]. This system of equations was solved by means of MATLAB built-in algorithm based on the 4th-order Runge–Kutta method. The unknown parameters σ_{36} - excited-state photoionization cross-section of Pr^{3+} ions, σ_{45}- ground state cross-section of Ce^{3+} ions at 266 nm and K- energy transfer coefficient from the ion Pr^{3+} ions to Ce^{3+} ions were determined by means of direct Hooke–Jeeves search and equal, respectively, to $1*10^{-18}$ cm^2, $1*10^{-21}$ cm^2 and $5.2*10^7$, $2.19*10^7$, $1.25*10^7$ c^{-1} for crystals with concentrations C_{Ce} = 0.47 at%, C_{Pr} = 0,24 at%; C_{Ce} = 0.38 at%, C_{Pr} = 0,28 at%; C_{Ce} = 0.16 at%, C_{Pr} = 0,38 at%, respectively. These results are in good agreement with the data of ref.[3] and above mentioned kinetic measurements.

Numerical simulation of optical gain on the 5d-4f transitions of Ce^{3+} ions in samples were carried out and the up-conversion pump optimal conditions for further pump-probe experiments were found. It were established that to realize optical gain coefficient more than 1.2 the all of the pumping beams should be π polarized and their radiation density should be more than 0.8 J/cm2. The optimal delay between the first (595 nm) and second (266 nm) pumping pulse should be about 15 ns.

5. Conclusion

Here the key parameters of processes provided up-conversion pumping of 5d-states of Ce^{3+} ions in Ce^{3+}, Pr^{3+}:$LiY_{0.3}Lu_{0.7}F_4$ mixed crystals via $^1D_2$ level of Pr^{3+} ions were estimated for the first time. The optimal up-conversion pump conditions were determined. The prospects of this material as the first up-convertionally pumped UV active media were demonstrated.

Acknowledgements

This work was funded by the subsidy allocated to Kazan Federal University for the project part of the state assignment in the sphere of scientific activities.

6.References

[1] Semashko V V 2012 *Active media of ultraviolet solid-state lasers. Prospects, obstacles and advances* (Saarbrucken: Palmarium Academic Publishing)
[2] Semashko V V, Joubert M F, Descroix E, et.al. 2000 *Proc. of SPIE.* **4061** 306–316
[3] Nicolas S, Descroix E, Joubert M F, et.al. 2003 *Opt.Mat.* **22** 139-146
[4] Nizamutdinov A S, Semashko V V, Naumov A K, et.al. 2008 *Phys. of Solid-State* **50** 1648-1651
[5] Gorieva V G, Semashko V V, Korableva S L, et.al. 2013 *Uchenie zapiski Kazanskogo universiteta* **155** 52 – 58
[6] Nikolas S, Descroix E, Guyot Y, et.al. 2001 *Opt. Mat.* 16
[7] Lim K-S, Hamilton D S 1989 *Opt. Soc. Am. B.* **6** 1401-1406
[8] Nicolas S, Descroix E, Guyot Y, et.al. *Opt. Mat* **16** 233-242
[9] Nizamutdinov A S, Marisov M A, Semashko V V, et.al. 2005 *Phys. of Solid State.* **47** 1460-1462
[10] Pavlov V V, Marisov M A, Semashko V V, at.al. 2013 *J. of Lum.* **133** 73-76

Toxicity of laser irradiated photoactive fluoride PrF_3 nanoparticles toward bacteria

M S Pudovkin, S L Korableva, A O Krasheninnicova, A S Nizamutdinov, V V Semashko, P V Zelenihin, E. M. Alakshin, T A Nevzorova

Kazan Federal University, 420008, Kremlevskaja str., 18, Kazan, Russian Federation

E-mail: jaz7778@list.ru

Abstract. The article is devoted to exploration of biological effects of crystalline PrF_3 nanoparticles toward *Salmonella typhimurium* TA 98 bacteria under the laser irradiation. Obtained results show bactericidal activity of PrF_3 nanoparticles and optimal parameters of laser irradiation (power of laser irradiation, wavelength, diameter of the laser spoil, and exposure time) have been found under which the effects of bactericidal activity become the most significant. Survival of bacterial cells under laser irradiation with wavelength 532 nm in colloidal solution of PrF_3 nanoparticles was 39%, 34%, 20% for exposure times 5 minutes, 15 minutes and 30 minutes, correspondingly.

1. Introduction

Nanosized materials are used in different application of industry. Because of the unique physical and chemical properties of nanoparticles they can be used in a broad range of biological applications. Several last researches [1-4] have reported that photoactive nanoparticles could cause oxidative stress and they can be activated by UV and visible irradiation generating reactive oxygen species (ROS), such as hydroxyl radical OH· and superoxide anion $O_2^{\cdot}$. In consequence, the photoactive nanoparticles can be applied in waste water treatments as bactericidal medium, in environmental remediation, organic contaminants mineralization, and cancer treatment [5, 6]. There are three traditional methods of cancer treatment: chemotherapy, surgery, and radiotherapy. Undoubtedly there are alternative methods of cancer treatment that can be used as independent methods or with traditional methods jointly. Therapy with photoactive nanoparticles is one of those alternative methods. The ROS, produced by photoactive nanoparticles, can damage cellular components like proteins and DNA and cause the death of treated cells. Selective oxidative stress is sometimes desirable and can be used therapeutically also. Semiconductor nanostructures TiO_2, ZnO with photocatalytic activity have already found a wide range of applications [1-4]. But these semiconductor nanostructures have relatively low photocatalytic activity. Besides, they could be activated by UV irradiation only. On the other hand there are very few reports devoted to biological effects of photoactive fluoride nanoparticles. In spite of this they are discussed as very promising nanosized materials for different biological applications because of their high absorption capacity in the wavelength range of 450 nm - 550 nm and high photoenergy conversion efficiency and they can be used in different biological applications, including cancer treatment [7]. Tests on procariotic sells is the first step of toxicity investigation according to traditional methodics.

The main aim of present investigation was to test hypothesis of toxicity of photoactive fluoride nanoparticles PrF_3 towards procariotic cells under laser irradiation.

2. Materials and methods

Team of Laboratory of Quantum Electronics and Radiospectroscopy named after S. A. Altshuller at Kazan Federal University has utilized a technology of synthesizing fluoride crystalline nanoparticles PrF_3 by chemical methods. The method is based on deposition from colloidal solution and hydrothermal reaction and is described in [8-12]. The HRTEM images of nanoparticles output are shown in the figure 1. The material used in present experiments consisted of crystalline nanoparticles with size about 30 nm. In the experiments we used CW lasers with wavelength 532 mn (green light) and 473 nm (blue light) with power 30 mW as activation irradiation sources. We have investigated toxicity of floride nanoparticles PrF_3 under the laser irradiation towards *S. typhimurium*. *S. typhimurium* TA 98 was cultivated in LB medium with 50 μg/ml ampicillin sodium for 18 h at 30^0C, then centrifuged (3000 g, 20 m) and the cell pellet was resuspended in physiological saline solution to the final concentration 10^6 cell/ml. 200 μl per well of bacterial cell suspension together with colloidal solution of PrF_3 nanoparticles was placed into white opaque 96-well microplates (PerkinElmer) and irradiated with lasers. Concentration of PrF_3 nanoparticles was kept at the level 0,1 mg/l in the well. Also control wells were kept without PrF_3 nanoparticles solution. After the irradiation cell suspension was diluted with physiological saline solution in 10^5 times and plated onto agarised LB medium with 50 μg/ml ampicillin sodium. After 2 days of incubation at 30^0C the number of colony-forming units was counted.

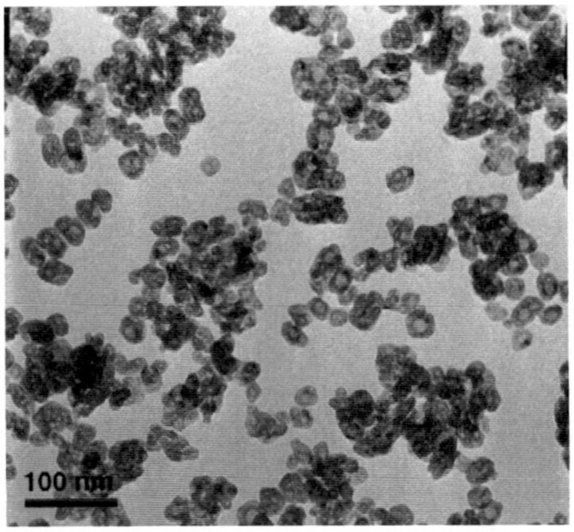

Figure 1. The HRTEM images of fluoride photoactive nanoparticles PrF_3 synthesized by means of hydrothermal reaction according to method described in [8].

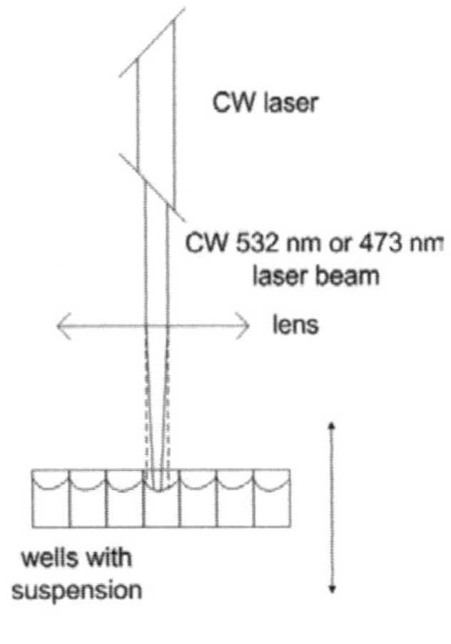

Figure 2. The experimental set up for photoactivated toxicity studies of PrF_3 nanoparticles.

The toxicity test consisted in exposure of the well with cells and nanoparticles to the laser radiation as it is illustrated at the experimental set-up scheme in the figure 2. We have made experiments for laser wavelengths 473 nm and 532 nm. Laser spot at the cell sample was formed by the lens with 7,5 cm focal length. We have investigated the influence of laser irradiation on the cells using three configuration of experimental set-up: well exposure under the laser irradiation without the lens (irradiation power density 637 mW/cm^2), the well in the visual focus of the lens (irradiation power density 1,3 W/cm^2), and the well behind the visual focus of the lens (irradiation power density 285 mW/cm^2). We have also investigated the influence of nanoparticles without the laser irradiation and the influence of laser irradiation without the nanoparticles.

3. Results and discussion

The first experiments at 473 nm and 532 nm wavelengths of irradiation have shown higher photoenergy conversion efficiency for the wells behind the visual focus of the lens under laser irradiation with wavelength 532 nm. Results of the toxicity tests have shown in Figure 3. Laser irradiation at 473 nm and 532 nm appears to be toxic itself with the higher effect of green light (survival index is 81 % for 473 nm light and 57 % for 532 nm light). Increasing the power density leads to increase of the toxic effect. Also the toxicity of PrF_3 nanoparticles is well pronounced. Presence of nanoparticles increases toxic effect in about 20 %. The toxic effect of irradiated nanoparticles showed some saturation as it slightly decreased with increasing the power density. The most significant toxic effect appeared to be for cases of 637 mW/cm2 of 532 nm laser light with 22 % of survival index and 285 mW/cm2 of 473 nm laser light with 62 % of survival index.

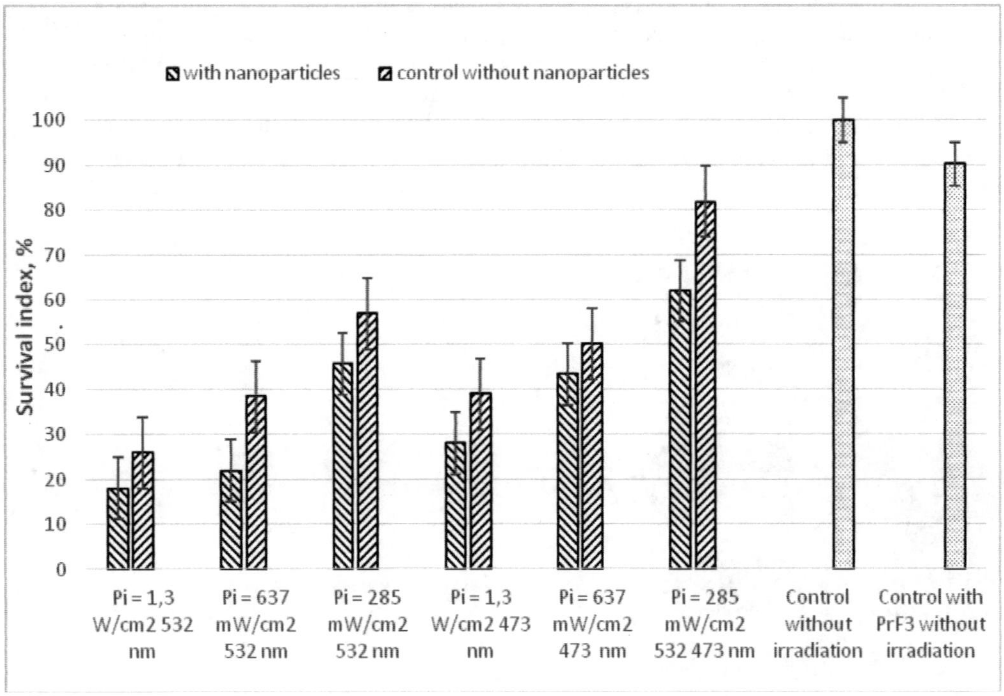

Figure 3. Survival of the *S. typhimurium* TA 98 cells with PrF_3 nanoparticles under various power density of irradiation.

In the second series of experiments we have investigated *S. typhimurium* TA 98 cells survival treated with nanoparticles at various exposure times of the laser irradiation. Three duration of exposure time: 5 minutes, 15 minutes, 30 minutes were chosen. It is seen from the diagram in Figure 4 that cells survival index decreases with increase of exposure. The number of CFU was 39%, 34%, 20% for exposure time 5 minutes, 15 minutes and 30 minutes compared with the control, correspondingly, under laser irradiation with wavelength 532 nm. The result of almost complete bacterial inactivation is apparently due to both effects of nanoparticles photoactivation and 532 nm irradiation itself. Behavior of survival for cells irradiated at wavelength 473 nm also speaks for bactericidal effect and appears to be exposure dependent but laser irradiation at this wavelength does not show good conversion efficiency. And treatment by PrF_3 nanoparticles increases the bactericidal effect.

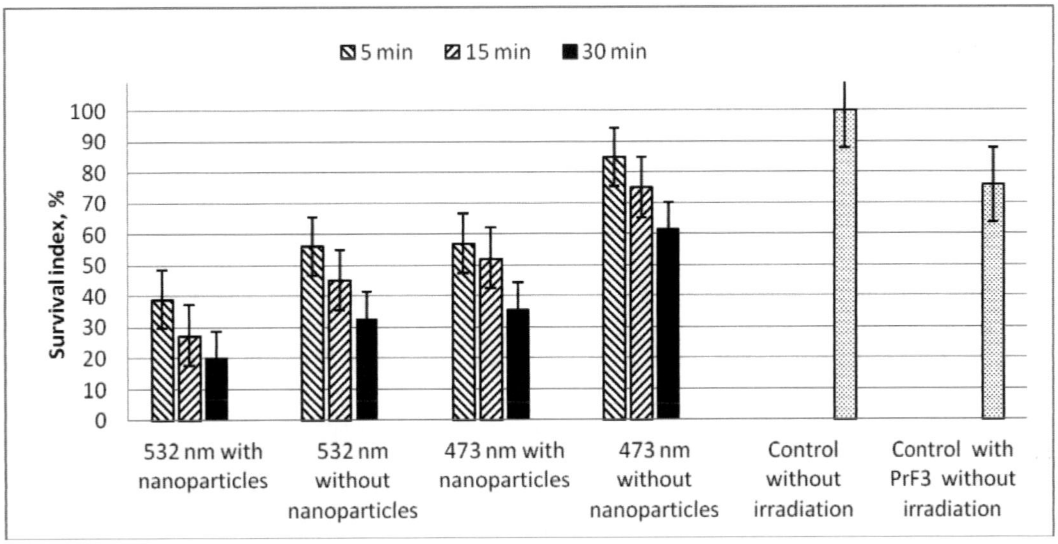

Figure 4. The influence of exposure time of laser irradiation at 532 nm and 473 nm on survival of *S. typhimurium* TA 98 cells treated with the PrF_3 nanoparticles for exposures 5, 15 and 30 min.

We have confirmed the hypothesis toxicity of photoactive fluoride nanoparticles PrF_3 toward *S. typhimurium* TA 98. These nanoparticles show photoactivated bactericidal activity of the same level as widely investigated materials on the basis of semiconductor nanoparticles (ZnO, TiO_2 [1-3,6]) which is about 18 % at 1,3 mW/cm^2 of 532 nm laser irradiation. Hypothetical mechanism of photoactivity could be also the same. It is known that H_2O inevitably appear at the surface and in the pores of fluoride nanoparticle prepared by hydrothermal method of growth [13-15]. When light is absorbed the photoexcited nanoparticle stores energy by charge separation and creating electron–hole pairs. These free charge carriers react with the adsorbed H_2O and thus formed hydroxyl radicals and holes (named as ROS) are responsible for bactericidal activity during irradiation of nanoparticle. Pure semiconductors usually exhibit low photoenergy conversion efficiency probably because of their relatively low charge separation efficiency and faster recombination of charge carriers [6]. It might be supposed that fluoride nanoparticles have the same mechanism of generating ROS. Laser-induced ROS production by photoactive nanoparticles may be one of the reasons for their toxicity. This effect is shown for titanium dioxide nanoparticles, which have bactericidal [16] and virucidal [17] activity because of ROS production induced by radiation. We suppose a similar mechanism of activation for PrF_3 nanoparticles.

4. Conclusion

In this work the photoactivated toxicity of crystalline PrF_3 nanoparticles towards *Salmonella typhimurium* TA 98 bacteria has been tested. The fluoride photoactive PrF_3 nanoparticles were synthesized in the laboratory by chemical method consisting in deposition from colloidal solution and hydrothermal reaction. The nanoparticles used were monocrystals with size of about 30 nm. We have confirmed the bactericidal effect of photoactive fluoride nanoparticles PrF_3 for *S. typhimurium* TA 98 under 532 nm laser light irradiation and it appeared to be significant (about 18 % of survival index) compared to semiconductor nanoparticles but the laser light have shown small bactericidal effect itself.

Acknowledgements
This work was funded by the subsidy allocated to Kazan Federal University for the project part of the state assignment in the sphere of scientific activities.

References
[1] Melissa F, José M, Amadeu M, Barata S 2014 *Science of the Total Environment* **470** 379-389
[2] Apalangya V, Rangari V, Tiimob B, Jeelani S 2014 *Applied Surface Science* **295** 108– 114
[3] Socorro O, Rodolfo Z, Blanca P 2013 *Journal of Hazardous Materials* **263** 28– 35
[4] Behnajady M, Hamed E 2013 *Chemical Engineering Journal* **228** 1207–1213
[5] Raghavendra A, Gunveen W, Niranjan D, Bonamali P 2013 *Environ Sci Pollut Res* **20** 6521–6530
[6] Weiwei H, Hyun-Kyung K, Wayne W, David M, John C, Jun-Jie Y 2014 *J. Am. Chem. Soc.* **136** 750−757
[7] Shizhu Chen, Cuimiao Zhang, Guang Jia, Jianlei Duan, ShuxiangWang, Jinchao Zhang, Size-dependent cytotoxicity of europium doped NaYF4 nanoparticles in endothelial cells 2014 *Materials Science and Engineering* **43** 330–342
[8] Lin Ma, Wei-Xiang Chen, Yi-Fan Zheng, Jie Zhao, Zhude Xu, Microwave-assisted hydrothermal synthesis and characterizations of PrF3 hollow nanoparticles 2007 *Materials Letters* **61** 2765–2768
[9] Alakshin E, Gabidullin B, Gubaidullin A, Klochkov A, Korableva S, Neklyudovac M, Sabitova A, Tagirov M 2011 *arXiv*.1104 0208
[10] Alakshin E M et al Spin kinetics of He-3 in contact with synthesized PrF3 nanoparticles 2011 *Journal of Low Temperature Physics* **162** 645-652
[11] Wang X, Li Y Rare-Earth-Compound Nanowires, Nanotubes, and Fullerene-Like Nanoparticles: Synthesis, Characterization, and Properties 2003 *Chemistry - A European Journal* **9 (22)** 5627-5635
[12] Alakshin E M et al Annealing of PrF3 Nanoparticles by Microwave Irradiation 2014 *Optics and Spectroscopy* **116** 721 -723
[13] Qibing Chang, Jian-er Zhou, Yongqing Wang, Guangyao Meng 2010 *Advanced Powder Technology* **21** 425–430
[14] Popov A.V., Orlovskii Yu.V., Vanetsev A.S., Gaitko O.M., Orlovskaya E.O., Sildos I. 2014 *Journal of Luminescence* **145** 774–778
[15] Alakshin E M, Blokhin D S , A M Sabitova, A V Klochkov, V V Klochkov, K Kono, Korableva S L , Tagirov M S 2012 *JETP Letters* **96** 181-183
[16] Melian J.A. et al The photocatalytic disinfection of urba wastewaters 2000 *Chemosphere* **41** 323–327
[17] Watts R J et al Photocatalytic inactivation of coliform bacteria and viruses in secondary wastewater effluent 1995 *Water Res* **29** 95–100

Journal of Physics: Conference Series **560** (2014) 012012 doi:10.1088/1742-6596/560/1/012012

Up-conversion luminescence of $LaF_3:Pr^{3+}$ crystal

O A Morozov, A K Naumov, A V Lovchev and E Yu Tselishcheva

Kazan Federal University, 420008, Kremlevskaja str., 18, Kazan, Russian Federation

uh4rad@rambler.ru

Abstract. $LaF_3:Pr^{3+}$ crystal is proposed as a promising media for an infrared quantum counter in the mid-infrared spectrum range. In this paper it has been shown that this crystal can work as an infrared quantum counter in the range of spectra of about 3 μm at the temperatures 170 K and 210 K with the efficiency of about 5%.

1. Introduction

This paper is devoted to a problem of creating a solid-state infrared quantum counter (IRQC) based on $LaF_3:Pr^{3+}$ crystal. The topic of IRQC creation is relevant due to a number of circumstances. Currently there are number of monographs about the up-conversion luminescence of rare-earth ions in crystals, one of them for example is the work of Chukova [1], which summarizes the results of work on the issue until 1980. Auzel in his paper [2] summarized the results of a similar orientation up to 2004. In addition there are several works in which close to practice results on IRQC were obtained on $LaCl_3$ crystal with REI. Thus, $LaCl_3$ crystal, activated by REI, occupies a special place in the study of up-conversion IRQC. This is due to its low phonon energy [3]. Transmittance of the crystal in the infrared spectrum region increases and probability of a nonradiative relaxation between closely spaced energy levels reduces. In the paper [4] it was directly demonstrated that energy levels and selection rules for REI in the anhydrous salt $LaCl_3$ are best suited to implement IRQC. The article [5] disclosed the results of the IRQC realization on the basis of $LaCl_3:Pr^{3+}$ for 4.5 μm spectrum range. It would seem that $LaCl_3$ crystal is an ideal media for the problem of IRQC. However, it has a significant disadvantage – the crystal is extremely hygroscopic. This property restricts greatly of IRQC at its base. But LaF_3 crystal hasn`t got this disadvantage, and it has a phonon spectrum width close to $LaCl_3$ crystal - 350 and 260 cm^{-1}, respectively. LaF_3 crystal has a broad band of transparency up to 9 μm and a low phonon wing.

However, most works on IRQC based on LaF_3 crystal contains information only about the possibility of such counters creation, which is based on authors theoretical assumptions of a simple transport phenomena from $LaCl_3$ crystal to LaF_3 crystal. In other works authors have implemented IRQC, but only in the infrared region from 0.8 to 1.5 μm. But we are interested in the region of 2 - 6 μm. Thus, this question still remains open.

2. Samples and research technique

The samples of LaF_3 crystal were grown in the MRS laboratory, Institute of Physics, Kazan Federal University, using the method of directional solidification in graphite crucibles. Three samples of LaF_3 crystal, doped by Pr^{3+} ions with concentration 0.1, 0.3 and 1 at. % were grown. The samples were cylinders of 6 mm diameter and 20 mm length. On the side surfaces of the boule two flattened areas were polished. They were parallel to the axis of the crystal, as well as to each other. It allows us to perform experiments with π- and σ- polarized light.

Experimental setup for carrying out studies of up-conversion transitions properties in $LaF_3:Pr^{3+}$ crystal contained several blocks. A block of tunable dye laser for sample pumping. A block of probe IR light for illumination of a sample, which consisted of removable positionally dependent filters and IR source "globar". A block of temperature setpoint with a cryogenic nitrogen gas cooling system and a thermal insulated box with a sample holder and a temperature sensor. A block of luminescence registration with a CCD spectrometer. This setup allows us to record the luminescence spectra of the samples, as well as their temperature dependence from 77 K up to a room temperature.

2. Luminescence spectra of $LaF_3:Pr^{3+}$ crystal

During the experiments at a room temperature without IR illumination, it was found that the maximum luminescence intensity for $LaF_3:Pr^{3+}$ crystal excited by π-polarized radiation was observed at an excitation wavelength - 0.5985 μm (16708 cm^{-1}), and for σ-polarized – at a wavelength of - 0.5945 μm (16820 cm^{-1}). The luminescence spectra of the samples with different concentration under excitation by a σ-polarized laser beam with a wavelength of 0.5945 μm are shown in figure 1. From these spectra one can see that the luminescence is also caused by the transitions from levels lying higher than $^3P_0$, and occurs even in the absence of IR illumination. The mechanism of this up-conversion luminescence excitation was considered in details and described in [6].

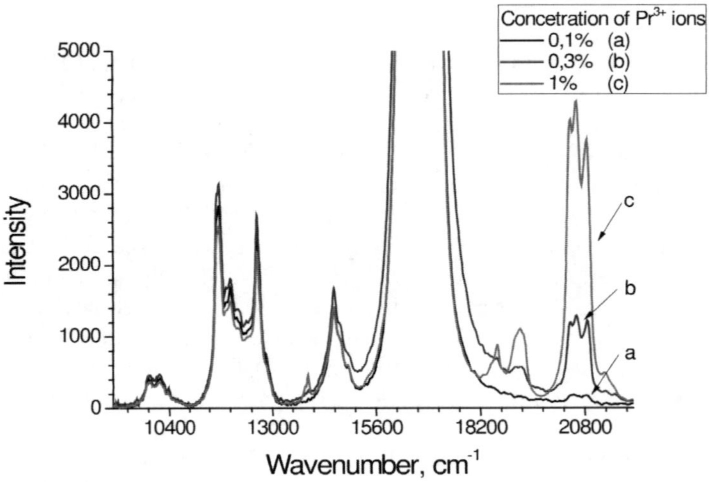

Figure 1. The luminescence spectra intensity of $LaF_3:Pr^{3+}$ samples with different concentration of Pr^{3+} ions

3. Properties of $LaF_3:Pr^{3+}$(1%) crystal as IRQC

The luminescence spectra of the samples in the temperature range from 140 to 280 K at a constant value of the σ-polarized pumping radiation energy in the visible range of the spectrum - 0.5945 μm with the illumination by IR radiation at 3 μm, and without it as well was recorded. When the temperature changes in the range from 140 to 280 K under illumination with IR radiation, we observed the luminescence intensity increase of individual lines. Figure 2 shows fragments of two luminescence spectrum: with IR illumination and without it. On the figure, one can see an increase in the intensity of up-conversion luminescence with IR illumination. Such luminescence response on IR illumination was stable in all experiments.

For obtained data analyze we constructed temperature dependences of relative intensity $(I_{IR}-I_0)/I_0$ versus a luminescence spectral lines of $LaF_3:Pr^{3+}$ (1%) crystal, which is shown in figure 3. Here I_0 - luminescence intensity of a pumped sample, I_{IR} - luminescence intensity of a pumped sample with additional illumination by IR radiation.

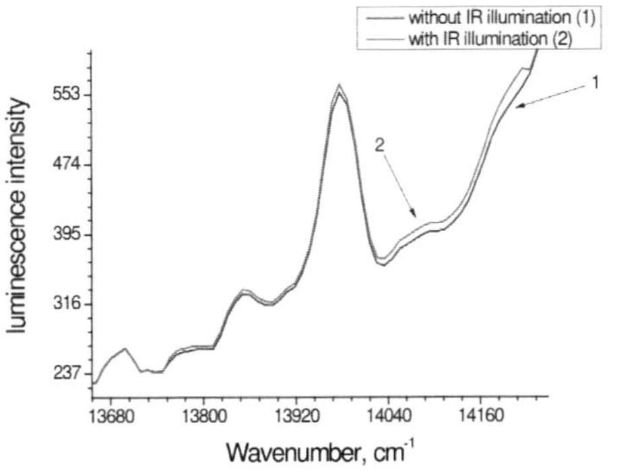

Figure 2. Luminescence spectra of $LaF_3:Pr^{3+}$ (1%) crystal 1 - without IR illumination. 2 - with IR illumination

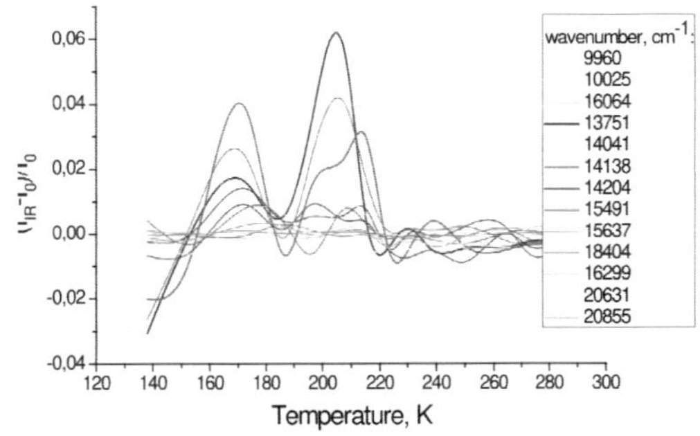

Figure 3. Dependencies of relative luminescence intensity $(I_{IR}-I_0)/I_0$ for a number of spectral lines for $LaF_3:Pr^{3+}$ (1%) crystal versus temperature

From the curves in the figure 3, one can see positive response to IR illumination from several spectral lines of luminescence, such as lines with wave numbers 13751 cm^{-1}, 14204 cm^{-1}, 14138 cm^{-1} and 20631 cm^{-1}. For other lines under illumination by an infrared radiation in the same conditions, there are no significant changes in their increase. It can also be seen from the figure that there are two temperature optima points, in which the response to IR illumination of some spectral lines has maxima. One of them corresponds to the temperature point of 170 K, and the other – 200 K.

4. Energy levels diagram of Pr^{3+} ions and electron transitions between them in LaF_3 crystal

For a better understanding of the processes occurring in $LaF_3:Pr^{3+}$ crystal under IR illumination it is necessary to take a look at the energy levels diagram of Pr^{3+} ions in this crystal and to transitions between them.

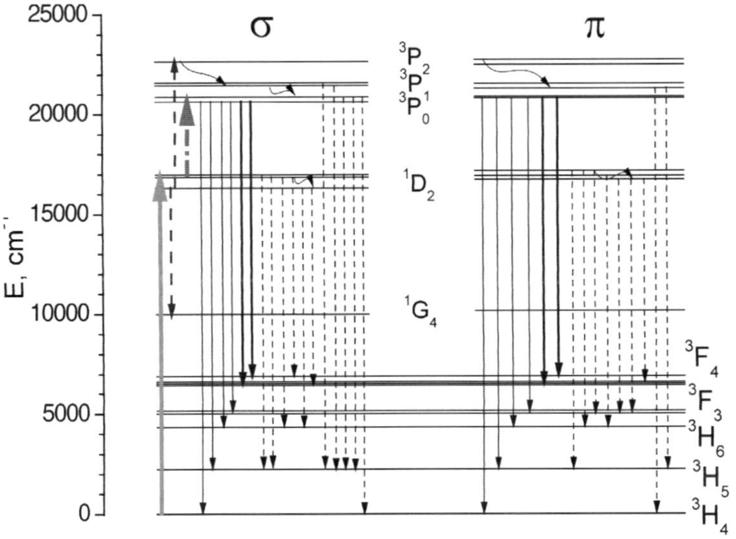

Figure 4. Energy levels diagram of Pr^{3+} ions and transitions between them in LaF_3 crystal

The interpretation of luminescence spectra (as transitions between states) with IR illumination shown on the energy levels diagram of Pr^{3+} ions (figure 4). The dash-dot arrow pointing upward on the diagram indicates the transition with absorption of IR quantum and excitation of Pr^{3+} ion from the excited (by radiation 0.5945 μm) state - solid arrow pointing up $^1D_2$ state to $^3P_0$ state. IR photon energy is approximately 3300 cm^{-1} or about 3 μm. It is evident that with IR illumination the intensity of individual lines of up-conversion luminescence corresponding to radiative transitions from $^3P_0$ level to the lower energy levels increases based on the plots shown in the figure 3. These transitions are marked on the diagram by solid arrows pointing downwards. Thus, in these experiments with $LaF_3:Pr^{3+}$ crystal pumped by 0.5945 μm radiation at low temperatures an increase in intensity of the individual lines of up-conversion luminescence with additional IR illumination with a wavelength of about 3 μm was observed and interpreted.

5. Conclusion

As a result of this work it has been shown that $LaF_3:Pr^{3+}$ crystal can be used as a quantum counter in the mid-infrared spectrum range. For this purpose the increment of up-conversion luminescence of pumped $LaF_3:Pr^{3+}$ crystal under IR illumination has been detected and interpreted.

The results are obtained under the following conditions: pumping was carried out with a pulsed laser with wavelength of 0.5945 μm and IR illumination of the crystal with a wavelength of 3 μm and power 10^{-7} W. Thus, at a low temperature 5% intensity increasing of the individual up-conversion luminescence lines was observed. Moreover, this effect occurred at two temperature optima 170 K and 210 K.

Increase of efficiency can be achieved by increasing of the interaction length of IR radiation with a pumped sample. For example, it can be done through the usage of crystalline optical fibers.

Acknowledgements

This work was funded by the subsidy allocated to Kazan Federal University for the project part of the state assignment in the sphere of scientific activities.

References

[1] Chukova Yu P 1980 *Russian book M.: Sovetskoye Radio* 193
[2] Auzel F 2004 *Chem. Rev.* **104** 139
[3] Soga K, Wang W, Riman R E, Brown J B and Mikeska K R 2003 *Appl. Phys.* **93** 2946
[4] John F 1961 *J. Appl. Phys.* **32** 825
[5] Case W E, Chivian S and Reynolds R W 1979 *J. Appl. Phys.* **50** 4509
[6] Morozov O A, Naumov A K, Korableva S L and Tselishchev D I 2013 *Optics and Spectroscopy* **114** 827

Photoconductivity of $SrAlF_5$ crystals doped with Ce^{3+} ions

V V Pavlov, V V Semashko, A N Yunusova, M A Marisov

Kazan Federal University, 420008, Kremlevskaja str., 18, Kazan, Russian Federation

E-mail: Vitaly.V.Pavlov@gmail.com

Abstract. Temporal behaviour of the complex permittivity of Ce:$SrAlF_5$ and Ce,Yb:$SrAlF_5$ crystals under laser irradiation in 240–280 nm spectral range was investigated by a microwave resonance technique. The photoconductivity spectrum of these crystals was detected. It was established that photoconductivity signal of the samples in 240-280 nm spectral range is caused by two processes: excited-state absorption of Ce^{3+} ions and photoionization of color centers.

1. Introduction

Until recently Ce-doped $SrAlF_5$ (SAF) crystal was considered as a promising active medium for UV lasers [1]. However stable laser action in this crystal has not been obtained till now. This is due to the fact that high energy UV radiation initiates the nonlinear photostimulated processes. Photostimulated processes caused by the impurity photoionization deteriorate spectral and energy characteristics of the vast majority of UV solid-state active media [2], and for Ce-doped SAF crystal these processes completely prevent any lasing. Therefore the study of impurity photoionization processes in activated crystals is an important task, since it allows us to determine the necessary conditions for lessening the harmful influence of nonlinear photodynamic processes. In addition it enables us to determine the positions of the energy states of the impurity centres in relation to the bands of the host.

It is known that one of the way to improve laser efficiency of Ce^{3+} doped fluoride crystals is the co-doping of the host crystal by ytterbium ions. In some Ce-doped fluoride crystals (for example Ce^{3+}:$LiYF_4$ and Ce^{3+}:$LiLuF_4$ crystals) the additional activation by Yb^{3+} ions engages a supplementary recombination channel for free carries and decreases the concentration of color centers [3]. The aim of this study is to detect photoconductivity spectrum of Ce:SAF and Ce,Yb:SAF crystals and to investigate the influence of ytterbium ions on the nonlinear photostimulated processes.

As the photostimulated processes significantly change complex permittivity of matter ($\varepsilon = \varepsilon_1 - j\varepsilon_2$) [4], dielectric spectroscopy measurements have been applied to achieve our aim. For instance, the variation of imaginary part of the permittivity of crystal $\delta\varepsilon_2$ can be caused by the appearance of free charge carriers generated as a result of one- or multi-photon impurity ionization, in other words, by photoconductivity effect. On the other hand, the variation of real part of the permittivity $\delta\varepsilon_1$ is associated with photodielectric effect. As opposed to the photoconductivity, photodielectric effect is caused by a change of dielectric polarization of the crystal under external electromagnetic action. Here electrons and holes, localized at impurity ions or traps, play the key role.

2. Materials

In this work Ce:SAF and Ce,Yb:SAF fluorite single crystals grown by means of the Bridgman technique have been studied. The concentrations of impurity ions (cerium and ytterbium ions) were

0.5 at% in the melt. It has been obtained earlier that during growth process ytterbium ions incorporate into the crystal host in trivalent and divalent states and the absorption bands of corresponding Yb^{2+} ions are overlapped with the absorption bands of Ce^{3+} ions in UV spectral region [5]. All crystalline samples were polished and shaped as parallelepipeds with the size of 1.5 × 1.5 × 1 mm. The size of the samples is limited to the experimental technique used in this work.

3. Experimental technique

Microwave resonant technique is one of the most informative methods of dielectric spectroscopy for studying the characteristics of photostimulated processes in rare-earth doped crystals [6]. This technique allows study of the change of both parts of the complex permittivity of matter undergoing different external influences (optical irradiation particularly) and registration photoconductivity and photodielectric effect separately. The microwave technique as compared with the conventional photoconductivity measurements with blocking electrodes has a number of advantages. Firstly, photoconductivity measurements in the microwave cavity are a contactless. When excluding the metal blocking electrodes we prevent the sample from distortion of energy bands and mutual displacement of the Fermi level and energy levels of the impurity ions in the investigated material in the vicinity of metal-dielectric contact, which could appreciably impact on research results. Secondly, the transient time of electromagnetic field in the cavity resonator is small enough to provide nanosecond temporal resolution (which is impossible via conventional method) and, therefore, it enables us to investigate rapid dynamics of the photostimulated processes.

The microwave part of experimental set-up is shown in figure 1 and has the signal and reference channels. The microwave source was a Gunn diode generator (G) that has the operating frequency of 35.4 GHz and the microwave power of about 70 mW. In the signal channel the microwave power passes through circulator (C) to the tunable rectangular resonator (R) where sample is placed in the antinode of the electric field. At first, the resonator is tuned on the operating frequency of Gunn diode by adjusting the volume. Then the sample is irradiated by the pulsed laser radiation that leads to the change of its permittivity. This in turn causes the change of microwave power reflected from the resonator and resulting change is registered by means of a quadrature balanced mixer (QBM). The mixer combines two balanced mixers that were operated at 90 degree phase shift. The use of quadrature balanced mixer and phase-shifter (Ph-Sh), located in the reference channel, allows us to register the variation of real and imaginary parts of the complex permittivity of sample separately. Detailed description of experimental set-up used in this work is given in [7, 8].

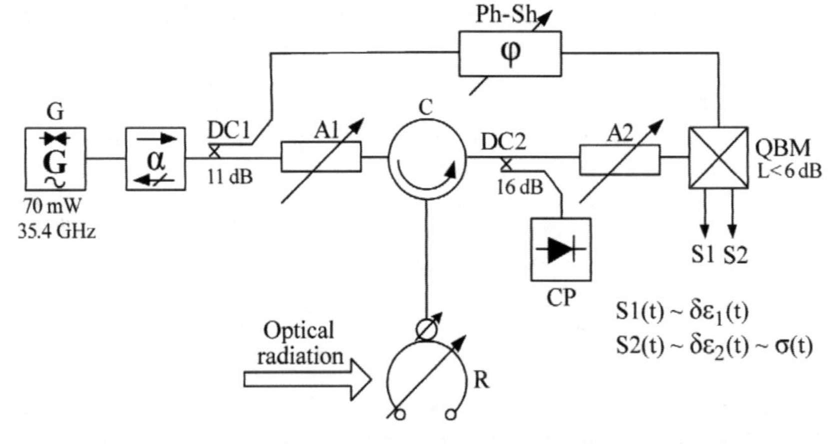

Figure 1. Schematic diagram of the microwave part of experimental set-up: G – Gunn diode; α – ferrite isolator; DC1 and DC2 – directional couplers; A1 and A2 – attenuators; C – circulator; R – tunable rectangular resonator; CP – microwave detector; Ph-Sh – phase-shifter; QBM – quadrature balanced mixer; S1 and S2 – photodielectric and photoconductivity signals, respectively.

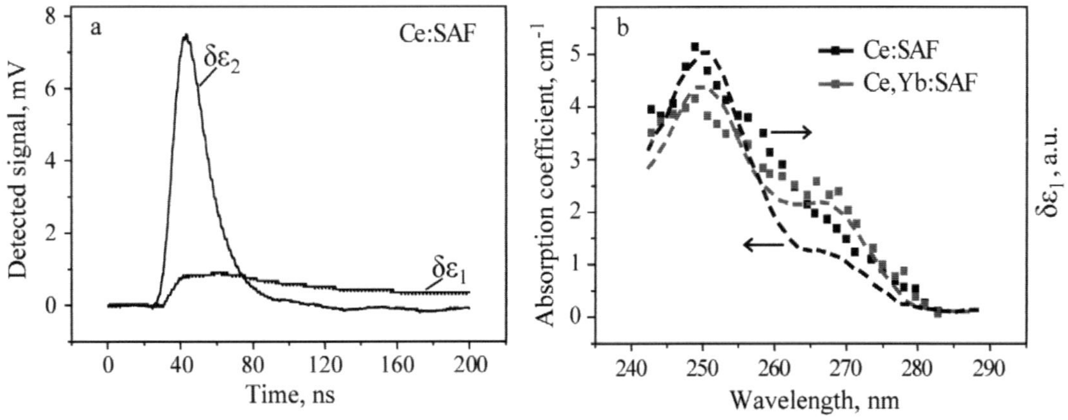

Figure 2. (a) Temporal behavior of real $\delta\varepsilon_1(t)$ and imaginary $\delta\varepsilon_2(t)$ parts of complex permittivity of Ce:SAF crystal excited by radiation at 266 nm; (b) absorption spectra (dashed line) [5] and spectral dependences of photodielectric signal $\delta\varepsilon_1(\lambda)$ (dots) for the Ce:SAF (black line) and Ce,Yb:SAF (grey line) crystals.

It should be noted that the transient time of electromagnetic field decay in the microwave cavity is about 2 ns. Taking into account 250 MHz-bandwidth of the microwave mixer and 200MHz bandwidth of the digital oscilloscope, which was used for a signal registration, the time constant of the measuring system enabled us to investigate transient responses of the dielectric permittivity of the crystals with about 5 ns temporal resolution.

The studied crystals were excited by radiation of the third harmonic of tunable Al_2O_3:Ti laser in spectral range from 240 to 280 nm. This spectral range corresponds to the 4f–5d transitions of the Ce^{3+} ions. Pulse duration and pulse repetition rate of the excitation radiation were less than 10 ns and 10 Hz, respectively. All experiments were performed at room temperature.

4. Experimental results and discussion

Timing performances of registered signal, corresponding to the variations of real and imaginary parts of the complex permittivity of sample, do not depend on sample and excitation wavelength. The temporal behaviour of the complex permittivity of Ce:SAF crystal is shown in figure 2a. The photoconductivity signal pulse form $\delta\varepsilon_2(t)$ for the studied crystals is characterized by the decay time of about 7 ns and almost coincides with one of the excitation pulse. That indicates the average lifetime of free charge carriers is less than 7 ns. The decay time of photodielectric signal $\delta\varepsilon_1(t)$, which characterizes the recovery time of dielectric polarization of crystal after pulse irradiation, appeared to be about 100 ns.

Normalized dependences of the amplitude of photodielectric signal versus excitation wavelength $\delta\varepsilon_1(\lambda)$ are shown in figure 2b. As it can be seen from figure 2b, the obtained spectra of photodielectric effect are similar to the ground state absorption spectra of Ce^{3+} ions. This means that interconfigurational 5d-4f transitions of the rare-earth ions significantly change dielectric polarization of crystal. Distinction between the lifetime of the excited 5d-state of Ce^{3+} ions, which is about 30 ns [9], and the decay time of photodielectric signal can be explained by the fact that the dielectric polarization of crystal cannot be changed instantaneously in response to an applied electromagnetic field.

Spectral dependence of the imaginary part of dielectric constant of the samples $\delta\varepsilon_2(\lambda)$ is shown in figure 3a. The photoconductivity spectrum of Ce:SAF crystal revealed a band with a maximum at 260 nm wavelength, which is absent for double-doped Ce,Yb:SAF crystal. Linear dependence of the photoconductivity signal for Ce:SAF crystal on the energy density of excitation radiation at 260 nm wavelength suggests that this band is essentially caused by one-photon ionization process.

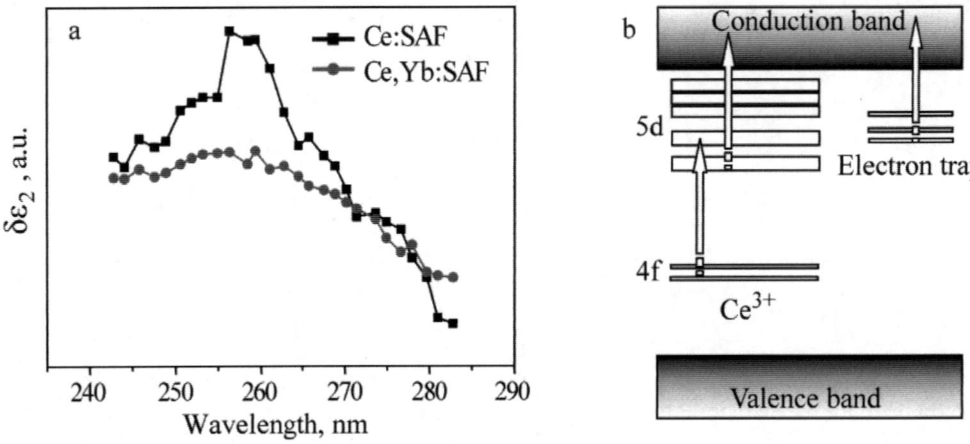

Figure 3. (a) Spectral dependences of photoconductivity signal $\delta\varepsilon_2(\lambda)$ for the Ce:SAF (black line) and Ce,Yb:SAF (grey line) crystals; (b) Interpretation of photoconductivity spectrum in 240-280 nm spectral range for Ce:SAF and Ce,Yb:SAF crystals.

As photoionization of Ce^{3+} ions in SAF crystal at the wavelength of 260 nm cannot be the result of one-photon absorption, the only possible interpretation of this 260 nm-band is photoionization of color centers. This confirms the fact that the additional activation by Yb^{3+} ions lowers the concentration of color centers. Therefore we can interpret the obtained photoconductivity spectrum of Ce:SAF and Ce,Yb:SAF crystals in 240-280 nm spectral range as superposition of two processes: excited-state absorption of Ce^{3+} ions and photoionization of color centers (figure 3b).

Acknowledgements
This work was funded by the subsidy allocated to Kazan Federal University for the project part of the state assignment in the sphere of scientific activities.

References

[1] Dubinskii M A, Schepler K L, Semashko V V, Abdulsabirov R Y, Korableva S L, Naumov A K 1998 *J. Mod. Opt.* **45** 221
[2] Pogatshnik G J, Hamilton D S 1987 *Phys. Rev.* B **36** 8251
[3] Nizamutdinov A S, Marisov M A, Semashko V V, Naumov A K, Abdulsabirov R Y, Korableva S L 2005 *Phys.Sol.State* **47** 1460
[4] Kazanskii S A, Rumyantsev D S, and Ryskin A I 2002 *Phys. Rev.* B **65** 165214
[5] Yunusova A N, Semashko V V, Safiullin G M, Nurtdinova L A, Pavlov V V, Marisov M A 2014 *J. Lum.* **145** 443
[6] Joubert M-F, Kazanskii S A, Guyot Y, Gacon J-C, Rivoire J Y, Pedrini C 2003 *Opt. Mat.* **24** 137
[7] Pavlov V V, Semashko V V, Rakhmatullin R M, Efimov V N, Korableva S L, Nurtdinova L A, Marisov M A, Gorieva V G 2013 *JETP Lett.* **97**, 1
[8] Pavlov V V, Semashko V V, Rakhmatullin R M, and Korableva S L 2014 *Opt. Spectr.* **116** 739
[9] Yunusova A N, Marisov M A, Semashko V V, Nurtdinova L A, Korableva S L 2012 *Opt.Commun.* **285** 3832

Photodynamic processes in $LiCaAlF_6$:Ce^{3+} UV active medium

A I Galiev, V V Semashko, O R Akhtyamov, S A Shnaidman, M A Marisov and A A Shavelev

Kazan Federal University, 420008, Kremlevskaja str., 18, Kazan, Russian Federation

E-mail: Almaz.Galiev.Isl@gmail.com

Abstract. The objectives of this paper are experimental studies of pump-induced effects in $LiCaAlF_6$: Ce^{3+} single crystals and computer model elaboration and appropriated software package engineering. The elaborated experimental technique and software allow either to calculate the nonlinear absorption/gain characteristics of the active medium on the basis of known parameters, or to find its previously unknown parameters from pump-probe experimental dependences.

1. Introduction

One of the key problems of quantum electronics is prospects estimation of using new materials as laser media. Existing methods of traditional optical spectroscopy are often cannot give a definite answer to this question, because during laser experiments the substance undergo much more intense impacts. Intense optical pumping is often accompanied by additional effects associated with the appearance of large electronic populations at the excited states of the investigated medium that may impede, impair or even completely exclude the possibility of laser action [1]. One of the effective ways to assess the real "laser potential" of new media are "pump-probe" spectroscopy methods, which suggest creation of conditions for the excitation of the material close to the conditions of the laser tests.The most informative studies are absorption/transmission coefficient investigation dependent on intensity and wavelength of exposure radiation. The specific role has interpretation methodology of the experimental data. Thus it is possible to fully characterize the active medium, quantify its prospects of realization, energetic and tunability properties of laser based on it.

2. Single-pass amplifier model

Experimental technique and samples preparations are described at our previous article [2]. By means of pump-probe spectroscopy absorption and gain coefficients of $LiCaAlF_6$: Ce^{3+} crystals have been registered in spectral ranges correspondent to the 4f $\leftrightarrow$ 5d transitions of Ce^{3+} ions versus the intensity of the pump and probe radiation. Interpretation of these dependencies is realized using a hypothetic model of photodynamic processes described in [1] and shown in Fig. 1.

The model included four level scheme of laser oscillator (levels 1-4) supplemented with transitions from excited states to the states localized in the conduction band of the crystal host (cross-sections σ_{35} at the wavelength of 266 (Pump) and 290 (Probe) nm) which result in generation of electrons and holes and photoconductivity, recombination processes (P_{51} and P_{54} probabilities) as one of the type of free charges relaxation channels and the processes of formation of color centers (P_{56} probability) as another one.

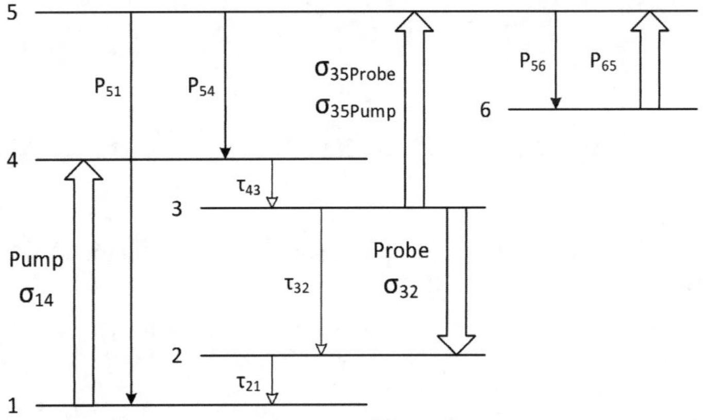

Figure 1. Stochastic model of photodynamic processes in $LiCaAlF_6:Ce^{3+}$ active media

The system also includes destruction of color centers channel both due to thermalization (P_{65} probability) and due to absorption of pump and/or laser radiations (cross-sections σ_{65} at the wavelength of 266 and 290 nm, respectively). The system of kinetic equations (1) corresponding to the model are following:

$$\frac{d}{dt}n_1 = -Pump(t)\cdot\sigma_{14}\cdot n_1 + \frac{n_2}{\tau_{21}} + Pump(t)\cdot\sigma_{14}\cdot n_4 + P_{51}\cdot n_5$$

$$\frac{d}{dt}n_2 = -\left[Probe(t)\cdot\sigma_{32} + \frac{1}{\tau_2}\right]\cdot n_2 + \left[Probe(t)\cdot\sigma_{32} + \frac{1}{\tau_3}\right]\cdot n_3$$

$$\frac{d}{dt}n_3 = Probe(t)\cdot\sigma_{32}\cdot n_2 + \frac{n_4}{\tau_4} - \left[Probe(t)\cdot\sigma_{35probe} + Probe(t)\cdot\sigma_{32} + Pump(t)\cdot\sigma_{35pump} + \frac{1}{\tau_3}\right]\cdot n_3$$

$$\frac{d}{dt}n_4 = Pump(t)\cdot\sigma_{14}\cdot n_1 - \left[Pump(t)\cdot\sigma_{14} + \frac{1}{\tau_4}\right]\cdot n_4 + P_{54}\cdot n_5 \quad (1)$$

$$\frac{d}{dt}n_5 = \left[Probe(t)\cdot\sigma_{35probe} + Pump(t)\cdot\sigma_{35}\right]\cdot n_3 - \left[P_{54} + P_{51} + P_{56}\right]\cdot n_5 + \left[Probe(t)\cdot\sigma_{65probe} + Pump(t)\cdot\sigma_{65pump} + \frac{1}{\tau_6}\right]\cdot n_6$$

$$\frac{d}{dt}n_6 = P_{56}\cdot n_3 - \left[Probe(t)\cdot\sigma_{65probe} + Pump(t)\cdot\sigma_{65pump} + \frac{1}{\tau_6}\right]\cdot n_6,$$

where σ_{ij} – transition cross-sections, P_{ij} – probability of processes, τ_i – lifetime of appropriated levels, $Probe(t), Pump(t)$ – probe and pump beam's photon density, n_i – appropriated level's population. The physical means of parameters of the system equations are clear from the Fig.1.

The finding solution of these equation systems are pump and probe radiations absorption or gain coefficients (2) which contents contribution of stimulated emission, ground-state, excited-state and color center absorptions:

$$k_{pump}(\mathrm{t}) = \sigma_{14} \cdot (n_1(t) - n_4(t)) + \sigma_{35pump} \cdot n_3(t) + \sigma_{65pump} \cdot n_6(t)$$
$$G_{probe}(t) = \sigma_{32} \cdot (n_3(t) - n_2(t)) - \sigma_{35probe} \cdot n_3(t) - \sigma_{65probe} \cdot n_6(t) \quad (2)$$

As the solution of the kinetic equation (1) is valid for a crystal layer of infinite small thickness, the balance equations need to be integrated in the spatial coordinates along the crystal. However, similar types system of integral-differential equations arises which is unsolvable analytically, and numerical methods are time-consuming. To simplify calculation procedure, the step-by-step approach is realized – the balk sample is considered as a stack of thin partitions which are absorbing or amplifying the pump and/or probe radiations accordingly by equations (1) and (2). Thus the beams characteristics from the output of previous partitions are using as input beams for the following ones. The number and thickness of partitions are dependent on the sample's length, parameter's values in the system equations (1), the measuring accuracy and a noise level. There is simple technique to estimate a minimal number of partitions N take into account the criteria (3) which demands that percentage change in the calculated value of the absorption or gain coefficients with the increasing N to N+1 should be less than experimental accuracy or noise level Δ:

$$k_{N+1} / k_N < \Delta \quad (3)$$

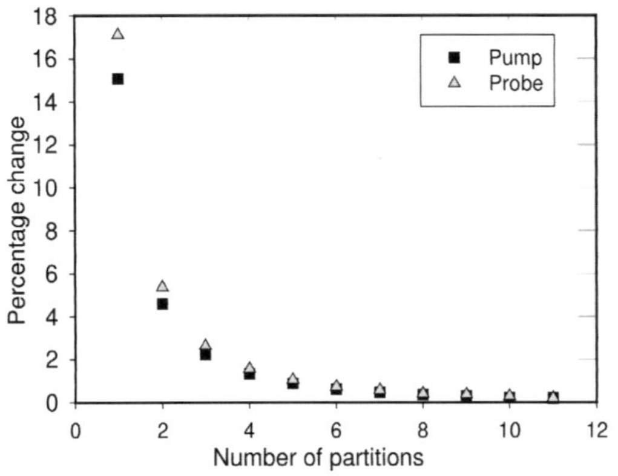

Figure 2. Determination of the optimal number of partitions.

For instance, it can be seen from the Fig.2 the optimal number of partitions is 8-10 for the typical samples length (~1 cm) and measuring accuracy (~1 %) in our experiments. To calculate the graphs on Fig.2 part of the model parameters were taken from the literature [3, 4] and spectral-kinetic studies of the samples. The remaining model parameters were the subject of an optimization procedure ("fitting") carried out in such a way to ensure a minimum residual between the experimentally determined and calculated data.

3. Results and discussion

The results of optimization procedure ("fitting") are shown in Fig. 3. and it are demonstrated a good agreement between experimental data and theoretical calculations.

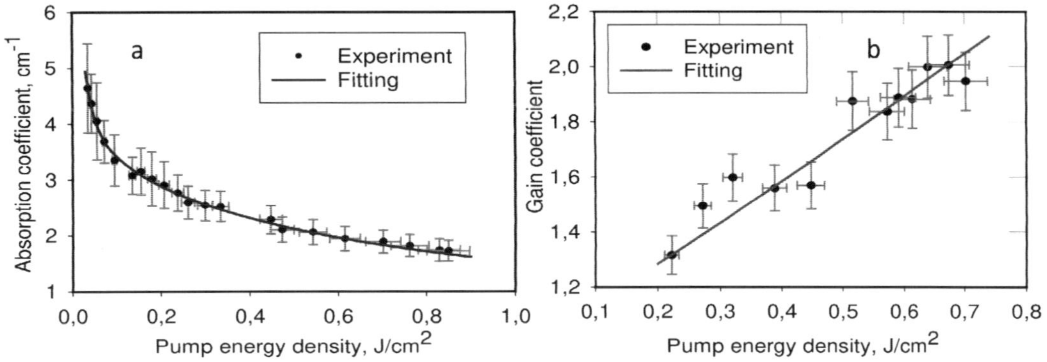

Figure 3. Absorption coefficient of pump radiation at λ=266 nm (a) and gain coefficient at λ=290 nm (b) depending on the pumping radiation energy density with the fitting results

Results are shown in the Table 1. Here are the estimated absorption cross-sections at the wavelength of 266 (σ_{35pump}) and 290 nm ($\sigma_{35probe}$) from Ce^{3+} ions 5d-excited-state and conduction band, the pump-induced color centers photoionization cross-sections at 266 (σ_{65pump}) and 290 ($\sigma_{65probe}$) nm, rates of recombination processes of free charge carriers to 5d- (P_{54}) and 4f-states (P_{51}) of Ce3+ ions and color center formation probabilities (P_{56}), concentration of Ce^{3+} ions and color centers.

Table 1. Parameters of photodynamic processes in $LiCaAlF_6$:Ce^{3+} resulting from computer simulation.

Absorption cross-sections, cm^2		Probabilities, s^{-1}		Concentration, cm^{-3}	
σ_{35pump}	$(1\pm0.3)\cdot10^{-18}$	P_{51}	$(3\pm0.4)\cdot10^{8}$	Ce^{3+} ions	$(7\pm1.2)\cdot10^{17}$
σ_{65pump}	$(3\pm0.5)\cdot10^{-17}$	P_{54}	$(9\pm0.4)\cdot10^{6}$	Color centers	$(3\pm1.5)\cdot10^{16}$
$\sigma_{35probe}$	$(3\pm0.5)\cdot10^{-19}$	P_{56}	$(5\pm0.4)\cdot10^{8}$		
$\sigma_{65probe}$	$(2\pm0.5)\cdot10^{-21}$				

As it can be seen from the Table 1, concentration of Ce^{3+} ions is much more than concentration of color centers, which confirms the lack of noticeable color center formation under the pumping conditions and explains high effective laser action in Ce:LiCAF active medium.

4. Conclusion

In this work for the first time the key parameters of photodynamical processes in Ce^{3+}: $LiCaAlF_6$ crystal active media were estimated. It are the 5d-excited-state photoionization cross-sections of Ce^{3+} ions and pump-induced color centers at pumping and probe beams wavelength together with rates of recombination processes of free charge carriers. Research results and developed software can be used for further design of ultrashort pulse laser based on the Ce^{3+}:$LiCaAlF_6$ single crystals, as well as for studies of photodynamic processes in other activated materials [5].

Acknowledgements

This work was funded by the subsidy allocated to Kazan Federal University for the project part of the state assignment in the sphere of scientific activities.

References

[1] Semashko V V 2005 *Physics of the Solid State* **47** 1507
[2] Galiev A I *et al* 2013 *J. Phys.: Conf. Series* **478** 012024
[3] Marshall C D 1994 *J. Opt. Soc. Am. B* **11** 2054
[4] Coutts D W, McGonigle A JS 2004 *IEEE Journal of Quantum Electronics* **40** 1430-1440
[5] Akhtyamov O R *et al* 2013 *Izv. VUZov. Fizika* **56** 39-42

The effect of chromophores concentration on the nonlinear optical activity of methacrylic copolymers with azochromophores in the side chain

M A Smirnov[1], A S Mukhtarov[1,2], N V Ivanova[1], T A Vakhonina[1], V V Semashko[3] and M Yu Balakina[1]

[1] A.E. Arbuzov Institute of Organic and Physical Chemistry, Kazan Scientific Center, Russian Academy of Sciences, 420088, Arbuzov str. 8, Kazan, Russia
[2] Kazan National Research Technological University, 420015, Karl Marx str., 68 Kazan, Russia
[3] Kazan Federal University, 420008, Kremlevskaya str., 18, Kazan, Russia

E-mail: macs-ne@mail.ru

Abstract. Quadratic nonlinear-optical characteristics of thin films based on methacrylic copolymers with of chromophore-containing monomers incorporated at various concentrations are measured by Second Harmonic Generation technique. Optimal chromophores content is obtained to be about 17 mol%, rather high values of nonlinear-optical coefficient, d_{33}, up to 60 pm/V, are determined.

1. Introduction

Nonlinear optical (NLO) characteristics of polymer materials originate from organic dipolar chromophores which are incorporated into the polymer by various ways [1,2]. Aggregation of these dipolar chromophores represents one of the governing factors in the development of polymer NLO materials limiting the accessible concentration of the chromophores in the polymer matrix; the aggregation was shown to occur even at moderate chromophores content [3,4]. Thus the determination of optimal chromophore concentration for a material to exhibit maximal quadratic NLO activity is a burning problem. This problem is realized by the researchers, however the systematic studies of the concentration dependence are rare in literature [3,5,6]. In these works different aspects of the development of electro-optic materials connected with the aggregation problem are considered: determination of the chromophores concentration at which maximal quadratic electric susceptibility, $\chi^{(2)}$, is realized [5]; relaxation stability of the poling order [3,6]; the ways for revealing chromophores aggregation [3,6]; the influence of pre-poling history on the effectiveness of chromophores orientation and $\chi^{(2)}$ value [6]; the effect of poling regimes on the $\chi^{(2)}$ value and temporal relaxation of the response [3,5,6]. Optimization of all mentioned factors may essentially enlarge NLO response and improve material properties. Conventionally in these papers the traditional polymer/dye systems are studied, both guest-host and polymers bearing chromophores in the side chain. Somewhat apart the researches stand where the new way of incorporation of chromophores is implemented, namely by means of noncovalent interactions between chromophores and polymer matrix [4,7]. The study of a tendency of aggregation in such systems depending on the nature of a polymer matrix is performed [4,7]. In all these studies the so-called azo-chromophores were used as NLO-active groups.

Methacrylic copolymers with incorporated organic NLO chromophore groups represent one of the classes of organic materials with quadratic NLO response to the applied electric field (see [1] and references therein). However, some problems still attract attention, in particular, the analysis of interplay between chromophores concentration and aggregation. Here we study the effect of chromophores concentration on the $\chi^{(2)}$ values by the example of methacrylic (co)polymers with azochromophores in the side chain, the concentration of chromophore-containing comonomer (MAZ) being 5 mol%, 8 mol%, 17 mol, 37 mol% and 43 mol% (see figure 1).

2. Experimental section

The synthesis of the studied methacrylic copolymers with various concentrations of MAZ monomers was carried out by the procedure described in [8].

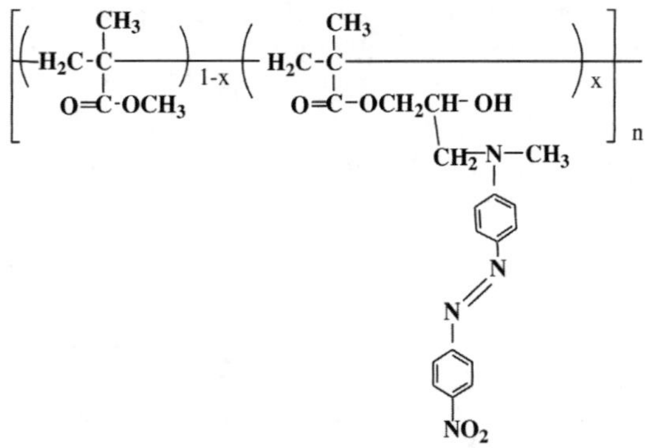

MMA-MAZ : x=0.05; 0.08; 0.1; 0.17;0.37; 0.43, n=60

Figure 1. Methacrylic copolymers with azochromophores in the side chain.

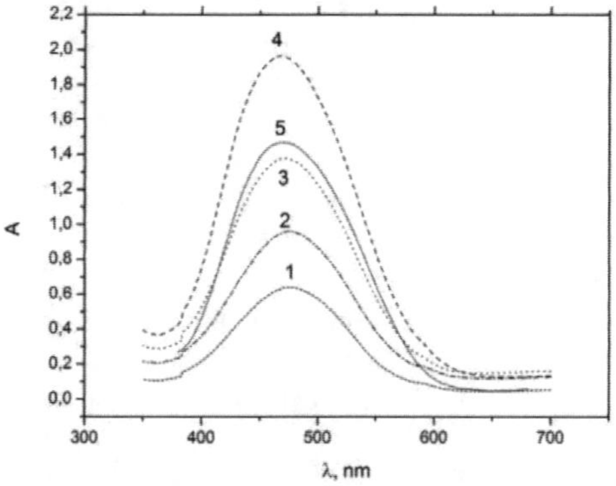

Figure 2. Absorption spectra in visible region of MMA-MAZ films with various molar content of chromophores: 5% (1), 8% (2),17% (3), 37% (4), 43% (5).

Polymer films were spin cast on glass plates from 7% solution of polymer in cyclohexanone (~5000 rpm during 60-90 sec); then the samples were kept in vacuum desiccator (drying oven) at room temperature for 10-16 hours followed by a soft bake at 50 °C during two hours. Film thickness was measured by AFM technique in tapping mode.

Since the material should be noncentrosymmetric in oreder to exhibit significant quadratic NLO activity, the polymer materials were converted into the so-called electret state with "frozen-in" macroscopic polarization appearing due to the orientation of dipole chromophores in the corona-discharge field [1,9]. This alignment was achieved in the electric field applied to the material heated to the temperature close to the glass transition temperature, T_g, the field was applied during ~20 min (t_p), after that the temperature was decreased with the field still on, when the room temperature was reached the field was switched off. Poling was carried out at the corona-triode setup, the corona discharge field voltage being 7 kV (positive corona), the distance from the tungsten needle electrode to the surface of the film being 1 cm. The orientation order parameter, characterizing the quality of chromophore poling, was controlled by UV/vis-spectroscopy by the change in absorbance values before (A_0) and after (/A) poling: $\eta=1- A/A_0$ [10]. Parameters characterizing poling regime are presented in table 1.

NLO coefficients of the studied polymers were measured by the Second Harmonic generation (SHG) technique (Nd:YAG laser, λ=1064 nm; 10 ns). The essence of the method is that we measure the intensity of SH-beam from the investigated polymer film and compare it with that of a standard. As a standard sample the alpha-quartz plate (X-cut) was used, the value of its nonlinear-optical coefficient d_{11} is equal to 0.3 pm/V [12]. Maximal intensity of the generated second harmonic beam was shown to be measured at the incident laser radiation angle of about 57°.

3. Results and Discussion

The optical absorption spectra of the obtained films are presented in figure 2 and demonstrate a small blue shift of absorption maximum for all the chromophores molar contents up to 37 mol%, this observation being consistent with those made in [4,11]. When chromophores concentration exceeds 37 mol% the absorption decreases; λ_{max} shows the red shift in this case, giving evidence to the chromophores aggregation [3]. The decrease of the d_{33} value, presented in table 2, starts already for the film with chromophores content of 37 mol% and in general is in agreement with this observation.

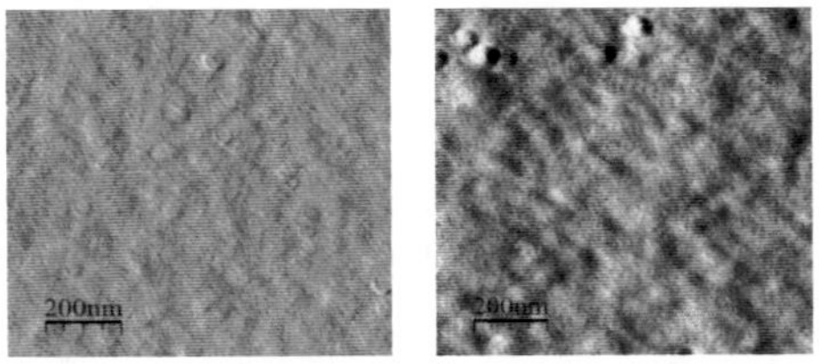

a **b**

Figure 3. The modification of film surface as a result of corona poling: **a** – before poling; **b** – after poling; AFM images.

Poling procedure results in the modification of the film surface (see figure 3); film roughness increases nearly twice.

The dependence of the obtained NLO coefficients d_{33} for the studied polymer films on the concentration of chromophores is presented in table 2. Increase of d_{33} value is observed up to level of concentration of 17 mol%, this being consistent with the formula for relation of molecular hyperpolarizability with quadratic electric susceptibility, obtained in the framework of the approach of non-interacting dipoles [2]. Maximum of d_{33} curve is observed near 17 mol %; then d_{33} decreases tending to saturation. Similar evidence was obtained in [5] for PMMA-DR1 side chain polymers, where optimal chromophores content was estimated to be 15 %. However, both in our case and in [5] films with limited number of various chromophores concentrations were studied, and the agreement between these conclusions could be considered reasonable.

The change of d_{33} values in table 2 is consistent with that obtained in [6] for guest-host films with pre-poling history. This fact gives hope to get higher nonlinearity when examining freshly-prepared films without pre-poling period during which spatial entanglements of polymer chains and head-to-tail arrangement of dipoles, hindering poling process, may take place [6]. Besides, optimization of poling protocol (T_p increase) may also serve to the achievement of this goal.

Table 1. Characteristics of the studied polymer films.

MAZ concentration (mol%)	Film thickness (nm)	λ_{max} (nm)	T_g (°C)	T_p (°C)	t_p (min)	Order parameter, η
5	700	475	126	125	20	0.04
8	570	475	128	112	20	0.33
17	410	472	131	125	20	0,32
37	410	468	138	127	20	0.36
43*	250	470	124	125	20	0.23

* - the procedure of film development differs in this case [8].

Table 2. The dependence of nonlinear optical coefficient d_{33} on molar concentrations of the chromophores in the side chain (MMA-MAZ). The numbers in brackets are standard deviations.

MAZ concentration (mol%)	d_{33} (pm/V)
5	21 (2)
8	46 (5)
17	60 (6)
37	53 (5)
43*	46 (5)

* - the procedure of film development differs in this case [8].

Thus, polymer materials based on methacrllic copolymers with different azochromophore molar contents, exhibiting relatively high values of NLO coefficients (up to 60 pm/V), were created.

Optimal concentration of chromophores in the studied methacrylic copolymers providing the highest value of the NLO coefficient is about 17 mol%.

Possible ways of enhancing quadratic nonlinearity are considered.

Acknowledgements
One of the authors (MAS) acknowledges financial support of the Academy of Sciences of the Republic of Tatarstan (grant №03-142т/Г2014); the work was partially funded by the subsidy allocated to Kazan Federal University for the state assignment in the sphere of scientific activities.

References

[1] Burland D M, Miller R D and Walsh C A 1994 *Chem. Rev.* **94** 31-75
[2] Kajzar F and Rau I 2013 *Rom. Rep. Phys.* **65** 663-80
[3] Rau I, Armatys P, Chollet P-A, Kajzar F, Bretonniere Y and Andraud C 2007 *Chem. Phys. Lett.* **442** 329-33
[4] Priimagi A, Cattaneo S, Ras R H A, Valkama S, Ikkala O and Kauranen M 2005 *Chem. Mater.* **17** 5798-02
[5] Kajzar F, Krupka O, Pawlik G, Mitus A and Rau I 2010 *Mol. Cryst. Liq. Cryst.* **522** 180-90
[6] Pawlik G, Rau I, Kajzar F and Mitus A C 2010 *Opt. Express* **18** 18793-04
[7] Priimagi A, Vapaavuori J, Rodriguez F J, Faul C F J, Heino M T, Ikkala O, Kauranen M and Kaivola M 2008 *Chem. Mater.* **20** 6358-63
[8] Vakhonina T A, Ivanova N V, Smirnov N N, Yakimansky A V, Balakina M Yu and Sinyashin O G 2014 *Mendeleev Commun.* **24** 138-39
[9] Giacometti J A and Oliveira Jr O N 1992 *IEEE Trans. Elect. Ins.* **27** 924-43
[10] Mortazavi M A, Knoesen A, Kowel S T, Higgins B G and Dienes A 1989 *J. Opt. Soc. Am. B* **6** 733-41
[11] Reyes-Esqueda J, Darracq B, García-Macedo J, Canva M, Blanchard-Desce M, Chaput F, Lahlil K, Boilot J P, Brun A and Lévy Y 2001 *Opt. Commun.* **198** 207-15
[12] Kaminskii A A, Bohaty L, Becker P, Eichler H J and Rhee H 2008 *Phys-Usp+* **51** 899-09

Laser characteristics of active medium $LiLu_{0.7}Y_{0.3}F_4:Ce^{3+}$ in ultra-short pulse mode

I I Farukhshin, A S Nizamutdinov, V V Semashko, S L Korableva
Kazan Federal University, 420008, Kremlevskaja str., 18, Kazan, Russian Federation

E-mail: burusha16@gmail.com

Abstract. We have obtained the single pulse laser oscillation with 400±10 ps pulse duration at 311 nm from $LiLu_{0.7}Y_{0.3}F_4:Ce^{3+}$ crystal with 6 ns laser pump at 289 nm. We expect this due to passive Q-switch by means of pump induced color centers which absorb laser radiation and determine the intensity discrimination of laser pulses with contrast of about 1,8.

1. Introduction

Today new technologies express demand on lasers oscillating in ultraviolet (UV) spectral range and having short pulse duration [1,2]. Main methods of obtaining UV laser radiation are nonlinear and parametric conversion of non UV laser radiation [1 – 5]. These methods can be technically difficult and it takes a lot of space. But on the other hand one can obtain UV lasing by use of fluoride crystals doped by Ce^{3+} ions as active media [6, 7]. Fluoride crystals doped by ions Ce^{3+} allow us to obtain short pulses with duration from several to tens of nanoseconds and even in subnanosecond time domain. To obtain shorter pulses we should use Q-switching or mode-locking [8, 9].

The aim of this work was obtaining laser pulses with ultra-short duration of UV spectral range and study of laser characteristics of active medium $LiLu_{0.7}Y_{0.3}F_4:Ce^{3+}$ (Ce:LLYF) in ultra-short pulse mode.

Among the trivalent lanthanide ions the ion of Ce^{3+} has minimal energy of 5d configuration with the value of about 50000 cm^{-1} [10, 11]. And ultraviolet laser radiation in Ce:LLYF is driven by stimulated emission resulting from interconfigurational 5d-4f transitions in Ce^{3+} ions excited by UV laser pump the same as in known homologous UV active medium $Ce^{3+}:LiLuF_4$ [6, 12, 13].

There are photodynamic processes in UV active medium resulting in formation of color centers [14]. Color centers in there turn absorb energy of laser radiation and cause losses in active medium [15, 16]. Level of losses due to color centers depends on amount of laser radiation inside the cavity and other factors (some of them are external additional irradiation and temperature) and doesn't remain constant during experiment [16, 17]. This gives an opportunity to utilize the color center losses for arrangement the modulation of Q-factor of the cavity or even mode-locking.

2. Experimental equipment

The object of the work is $LiLu_{0.7}Y_{0.3}F_4$ mixed crystal with 1 at. % of Ce^{3+} ions put in the melt. The sample was cut and polished according to side pumping scheme and lasing through Brewster windows. This sample was grown and prepared in Kazan (Volga region) Federal University in Quantum Electronics and Radiospectroscopy department.

The experimental setup for obtaining ultra-short pulses in UV band we have built is presented in Fig. 1. It contains 10 Hz pulsed UV solid state laser with double pass amplifier based on Ce^{3+}:$LiCaAlF_6$ active medium as pump source oscillating at 289 nm wavelength and 2 mJ energy level per pulse.

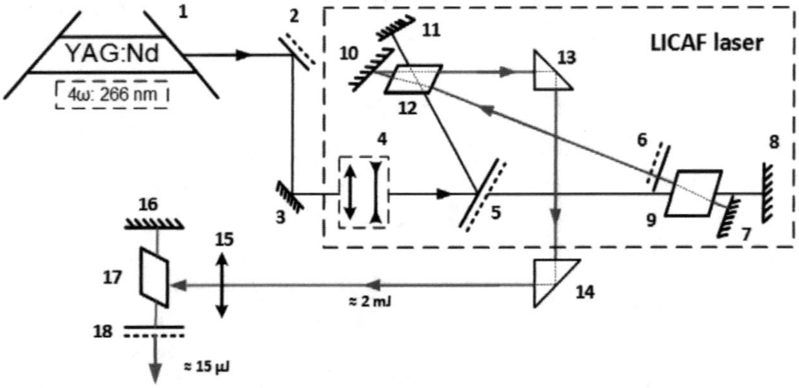

Figure 1. The experimental setup for the Ce:LLYF laser. Consists of: 1 – laser YAG: Nd; 2 - spectrum splitter; 3,7,8,10,11,16 – reflectors R=99,9%; 4 – telescope; 5 – 266 nm beam splitter R=30%; 6,18 – reflectors R=65%; 9, 12 – crystals Ce:LICAF; 13, 14 - rectangular prisms; 15 - converging lens; 17 – crystal Ce:LLYF.

The Ce:LLYF cavity was organized by two reflectors (99,9 % and 65 % reflection in the range 270 nm – 350 nm) and corresponded to Fabri-Perroit cavity with low Q-factor determined by small length (20 mm) and low output window reflection.

Pulse duration measurement was performed by Alphalas photodiodes with 50 ps rise time coupled to Tektronix oscilloscope with 3,5 GHz bandwidth.

3. Ultra-short pulse UV lasing

Laser oscillation from Ce:LLYF was obtained in pulse mode at the wavelength λ = 311 nm and with pulse duration t = (400±10) ps (Fig. 4b). Laser radiation of pumping was pulsed with pulse duration t = (6,375±0,025) ns (Fig. 4a).

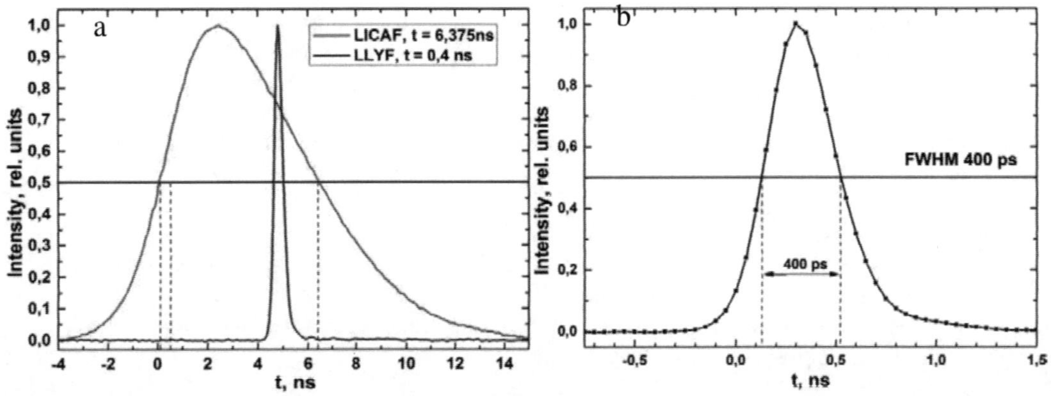

Figure 4. Temporal distribution of laser pulses obtained from Ce:LLYF (a) and Ce:LICAF (b)

The estimated photon lifetime in the cavity was τ_c = 281 ps. Pulse duration of Ce: LLYF laser radiation appeared to be greater than the photon lifetime in the cavity, so it means that we have multimode nature of the laser radiation. However, we see the single pulse, which tells us about possible Q-switching.

4. Dependence of the laser radiation energy on pump radiation energy.

To determine the cause of shortening the pulse duration we have studied the dependence of the output energy on the pump energy at different pulse repetition rates and the reflection coefficients of the output mirror of the cavity. The experiment consisted also in measuring the output energy sequentially with increase and decrease of the pump energy [16]. In result we see a hysteresis loop determined by color centers accumulation inside the active medium and decrease of its square with decrease of pulse repetition rate is characteristic due to finite lifetime of color centers.

Results of study of dependence of the laser radiation energy on pump radiation energy for different reflection coefficients of output mirrors have shown that reflection coefficient increase also leads to reduction of hysteresis loop square (see dependencies for 10 Hz repetition rate at Fig. 5 a and b), which is proportional to the number of color centers. Apparently lifetime of color centers is larger than time period between pump pulses and they do not dissolve to the time of next pump pulse and amount of color centers accumulated rises with increase of pulse repetition rate.

As it was investigated earlier [15, 18] increase of laser radiation intensity inside the cavity leads to bleaching of color centers. Thus for lower reflection coefficient 65 % of an output mirror an early saturation pronounces which is absent for 80 % reflection mirror. This is also an evidence of accumulation of color centers staying inside the active medium but experiencing photobleaching by higher laser radiation intensity for the case of 80 % reflection mirror.

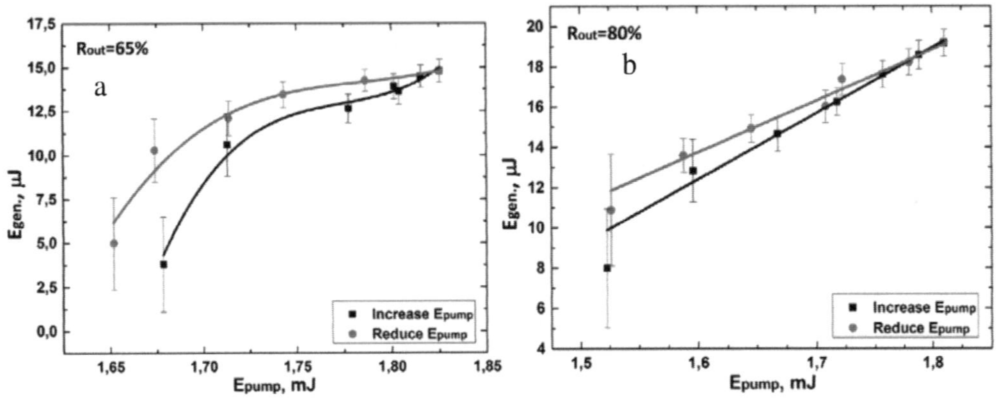

Figure 5. Dependence of Ce:LLYF laser output energy on pump radiation energy for mirrors with reflection coefficients R=65% (a) and R=80% (b) at 10 Hz pulse repetition rate

The photobleaching of color centers by laser radiation is also clearly seen from Fig. 6, where intracavity losses calculated from experimental results [18] are presented dependent on pump energy. Losses at higher energies of pump radiation where intensity of laser radiation is also high appear to be lower. And this is the principle of discrimination of laser pulses by means of saturable absorber. We have obtained the most significant contrast ratio for the highest pulse repetition rate in our experiments 10 Hz which corresponds to the case of accumulation of color centers inside the media.

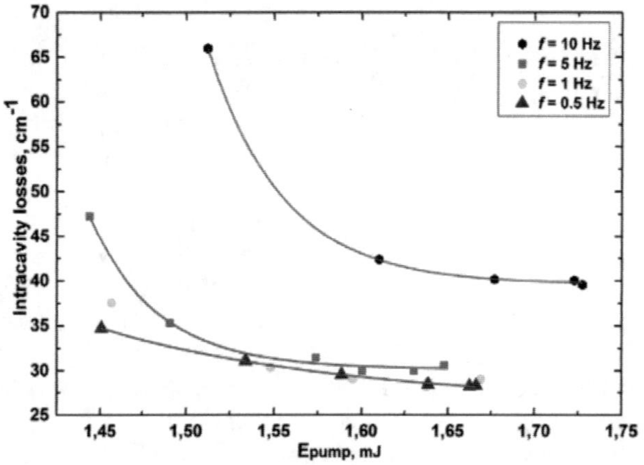

Figure 6. Intracavity losses for different frequencies.

5. Conclusion

Laser oscillation was obtained in active medium $Ce:LiLu_{0.7}Y_{0.3}F_4$ in a pulse mode at the wavelength $\lambda = 311$ nm as single pulse with duration t = (400±10) ps in the low Q-factor cavity. Pumping was performed at wavelength $\lambda = 289$ nm in pulsed mode with pulse duration t = (6,375±0,025) ns. Slope efficiency measured for 80 % reflection of output mirror in Fabri-Perroit nonselective cavity was 4 %. We expect that shortening of the pulse duration was achieved by passive Q-switch, which is caused by bleaching of color centers in the volume of the active medium absorbing the laser radiation. Thus measured contrast of losses level dependence on pump energy appeared to be 1,8 and methods of Q-switch control are change of pumping energy, pulse repetition rate or output mirror reflection.

Acknowledgements

This work was funded by the subsidy allocated to Kazan Federal University for the project part of the state assignment in the sphere of scientific activities.

References

[1] Saliminia A, Proulx A, Vallée R 2014 *Optics Communications* **333** 133-138
[2] Geernaert T, Becker M et. al. 2010 *J. LightwaveTechnol* .**28** 1459
[3] Beugin V, Bigot L et. al. 2006 *APPLIED OPTICS* **45**
[4] Zvorykin V D, Ionin, et. al. 2014 *Quantum Electronics* **44** 431-439
[5] Palacios A, Rivière, et. al. 2014 *Springer Series in Chemical Physics* **107** 25-48
[6] Dubinskii M A, Cefalas et. al.. 1992 *Optics Communications* **94 (1-3)** 115-118
[7] Dubinskii M A, V V Semashko et. al. 1993 *J. Mod. Opt.* **40** 1
[8] Sarukura N, Liu Z, et. al. 1995 *Optics Letters* 20 (6) 599-601
[9] Pham M H, Cadatal-Raduban M et. al. 2014 *Japanese Journal of Applied Physics* **53 (6)**, no. 062701
[10] Brewer L 1971 *JOSA* **61** 1666–1682
[11] Dieke G H, Crosswhite H M 1963 *Appl.Opt.* **2** 675–686
[12] Nizamutdinov A S, Semashko V V et. al. 2010 *JETP Letters* **91 (1)**, 21-23
[13] Nurtdinova L A, Semashko V V et. al. 2013 *JETP Letters* **96 (10)** 633-635
[14] Hamilton D S, Budger A B, Pinto A 1985 *Tunable solid state lasers* 80–90
[15] Lim K-S, Hamilton D S 1988 *Journal of Luminescence* **40&41** 319–320
[16] Semashko V V, Dubinskii et. al. 2002 *Proceedings of SPIE - The International Society for Optical Engineering* **4766** 119-126
[17] Lim K-S, Hamilton D S 1989 *J. Opt. Soc. Am. B.* **6** 1401–1406
[18] Semashko V V, Naumov A K, Nizamutdinov A S et. al. 2004 *SPIE Proc.* **5402** 421–429

The development of polymer laser-active media with improved performances

R Idrisov[1], V Serova[1], I Farukhshin[2], A Nizamutdinov[2], V Semashko[2], A Lovchev[2], A Naumov[2]

[1]Kazan National Research Technological University, K. Marks str., 420015 Kazan, Russian Federation

[2]Kazan (Volga region) Federal University, 420008, Kremlevskaya str., 18, Kazan, Russian Federation

E-mail: rav140988@gmail.com

Abstract. The influence of modification of organic glass by additives new derivatives of thiazole and thiourea on the spectral-fluorescent and lasing characteristics of the Rhodamine 6G laser dye has been investigated. For the sample modified by additive - N-(4-carbomethoxy-5-phenyl-thiazol-2-yl)-N'-phenylthiourea taken in amounts of 0.01 mol% the maximum rise of the absorption and fluorescence intensity to compare with the unmodified sample 1.3 and 2.3 times, respectively, was detected. The 8.3 times more lasing energy, and the 7.3 times higher laser photostability has been achieved.

1. Introduction

Optically transparent polymers and organic glasses based on methacrylates doped with laser dyes represent a new class of solid-state laser-active media (LAM). Research results of their properties are described in [1-3] and in other papers. The polymeric LAM successfully replaces liquid LAM - solutions of laser dyes. However, in order to broaden an area of practical application of laser devices with polymeric LAM it is necessary to improve their performances. It is well known that during the synthesis and subsequent treatment of the polymeric LAM the polymer matrix and incorporated dye undergo irreversible changes caused by free radical chain processes. As it is known [4,5], these processes lay in the basis of photochemical ageing and degradation of polymers. Therefore it is necessary to investigate new highly effective stabilizers.

In this paper new derivatives of thiazole and thiourea (DTT) were studied as modifying compounds (potential photostabilizers) synthesized in Institute of Organic and Physical Chemistry named after A.E. Arbuzov, Kazan scientific center of Russian Academy of Sciences. Namely these compounds are: N - (thiazol-2-yl)-N'-phenylthiourea (DTT-1) and N-(4-carbomethoxy-5-phenyl-thiazol-2-yl)-N'-phenylthiourea (DTT-2). In order to provide the comparison the commercially available N, N'-diphenylthiourea (DPTU) was also investigated as reference specimen. Structural formulas of all additives are shown in Figure 1.

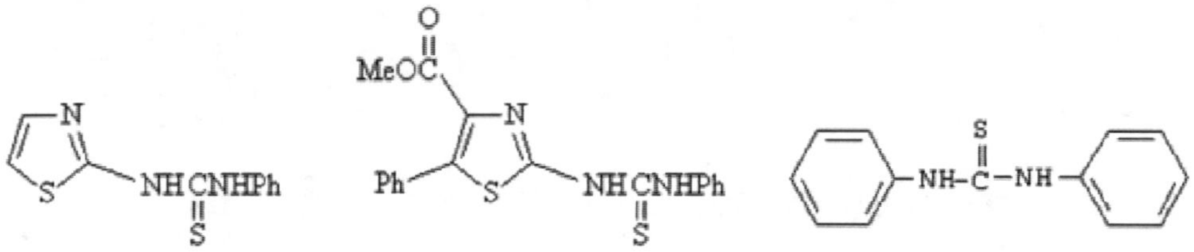

1 2 3

Figure 1. Structural formulas of modifying compounds: 1 is N - (thiazol-2-yl)-N'-phenylthiourea (DTT-1), 2 - N-(4-carbomethoxy-5-phenyl-thiazol-2-yl)-N'-phenylthiourea (DTT-2), 3 - N, N'-diphenylthiourea (DPTU)

Here we tested the modification of DTT, because of previously we have been established the ability of these compounds to inhibit (suppress) free-radical chain processes [6].

This work was aimed at the study of fluorescence spectral properties and laser characteristics of samples of organic glasses doped with Rhodamine 6G and DTT, and to elucidate the possibility of using them as LAM with improved performances.

2. Experimental

The objects of our research were prepared in the form of plates of organic glass doped with classical Rhodamine 6G laser dye –. The concentration of Rhodamine 6G in the initial mixture was $2{\cdot}10^{-4}$ $mol{\cdot}L^{-1}$ for the study of spectral-fluorescent properties and $4{,}57x10^{-4}$ $mol{\cdot}L^{-1}$ for laser tests. The thickness of the synthesized samples was 0.5 mm and 1.5 mm, respectively. Modifiers concentration have been varied between 0.01 - 0.10 mol. %.

In the work methylmethacrylate (MMA) and methacrylic acid (MAA) purified by vacuum distillation were used. Here are some physical properties of the compounds: MMA - refractive index 1.4130, density 0.943 $g \bullet cm^{-3}$, MAA - refractive index 1.4314, density 1.0153 $g \bullet cm^{-3}$.

DTT and Rhodamine 6G were dissolved in a mixture of MMA and MAA, and then their block radical-free copolymerization was carried out. The weight ratio MMA:MAA was equal to 90:10. Azobisisobutyronitrile (0.1 wt.%) as the initiator of reaction was used.

Registration of absorption spectra was carried out on a dual-beam scanning spectrophotometer «Lambda 35» (Perkin-Elmer, USA). To excite fluorescence in the samples UV solid-state Ce:LLF picosecond laser (Ultraviolet Solution LLC) was used. It provided single 400 ps laser pulses with pulse repetition rate 10 Hz at the wavelength $\lambda = 310$ nm. The fluorescence spectra were studied by Stellarnet spectrometer and fluorescence kinetics were registered by Alphalas UPD photodiodes with 50 ps rise time and 3.5 GHz bandwidth Tektronix DPO 7354 digital oscilloscope. The CW fluorescence spectra were also registered using a spectrofluorimeter Cary Eclipse (Varian, USA).

The scheme of experimental setup which used for study of the sample performances is presented on Fig.2. To achieve the laser oscillation from investigated samples and to study it photostability the second harmonic radiation of the Nd:YAG laser was applied. The study of the laser properties was realized in the flat-parallel cavity which formed by EC and OC couplers with quasilongitudinal scheme of lasing excitation. The resonator length was 70 mm, the angle between cavity axis and pump beam was about 7°. Measurement of every pulse falling energy, not absorbed energy and oscillation energy of a sample were carried out with photodiodes with have peak-detectors, Ph1 – Ph3 respectively. The parts of energy beam from every point of measurements were branched off to every photodiode with help of flat-parallel plates P. Measurement head MH allow of absolute energy measuring and it is used for calibration of photodiode Ph1-Ph3. For starting of measuring processes of the setup from every pulse of pumping was used the triggering photodiode TPh. Energy values from all of point measurements of the setup were drawn on the computer's monitor PC. Energy density of pumping sample was being changed by changing energy with using optical attenuator OA.

Degradation of laser oscillation in polymer active media were carried out in the cavity with highly reflecting coupler - EC (99,9 % reflection in the range 500 nm – 650 nm) and output coupler - OC with 0,6 % reflection around 600 nm. Energy density of the pump radiation was kept at the level about 0,142 J/cm^2.

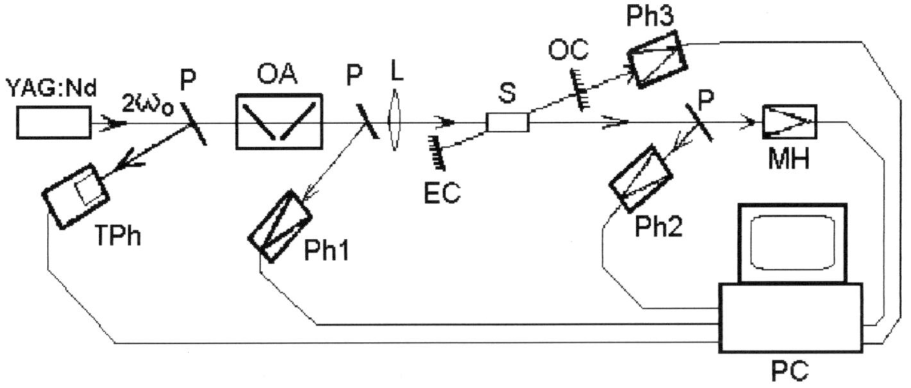

Figure 2. The scheme of experimental setup arrangement for sample oscillation study.YAG:Nd laser; P –flat-parallel plates; OA – optical attenuator; L – lens; S-sample; EC – end coupler; OC – output coupler; TPh – photo receiver for starting measurement process; Ph1-Ph3 photo diodes; MH – measurement head for absolute measurement energy value; PC – personal system.

3. Results and discussions

Figure 3 illustrates the influence of concentration of the modifying compound DTT-2 in the polymer matrix doped with laser dye on the fluorescence and absorption characteristics. Figure 3 shows that both absorption coefficient and fluorescence intensity of rhodamine 6G are increased in the modified polymer matrix at concentration of DTT-2 of 0.01 mol%. It is explained by an increasing of molecular stability of the dye in copolymerized system. That is, the introduction of a modifier has ensured safety of higher amount of optically active molecular centers of rhodamine 6G during synthesis process. However, with increasing concentration of DTT-2 in the polymer matrix the decreasing of fluorescence intensity of the dye (particularly at 0.10 mol %) is observed. It can be explained by reabsorption of fluorescence of rhodamine 6G by the modifying additive. Fluorescence of the dye occurs in the region λ = 530-650 nm, which is characteristic of the rhodamine 6G dye. Bands observed in the shortwave region of the spectrum - in the range of λ = 350-475 nm, apparently correspond to the intrinsic fluorescence of the polymer matrix.

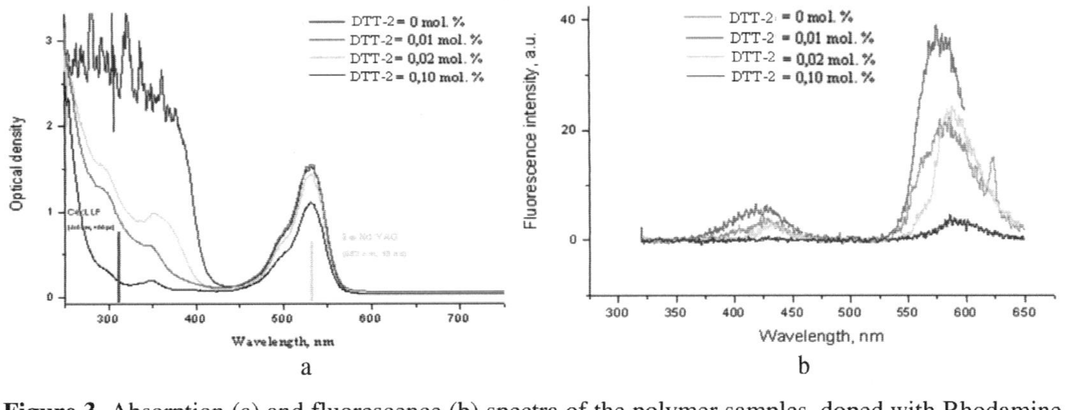

Figure 3. Absorption (a) and fluorescence (b) spectra of the polymer samples, doped with Rhodamine 6G and modified by DTT-2

Results of the study of spectral-fluorescence properties of the check (unmodified) sample, and the samples modified by additives DTT presented in Table 1. The table contains data characterizing the position of the maximum fluorescence (λ_{fl}) and fluorescence intensity values (I). The results of laser tests of doped polimer samples are also presented. These are presented as relative change in energy of generation – E_m/E and laser photostability of dye in a polymer matrix - N_m/N (E_m and E – output energy of laser radiation of the sample with the additive of the modifier and check one, respectively; N_m and N - number of pumping radiation pulses, that lead to half the height dropping of the output energy of laser action. In Table 1 the corresponding characteristics are also given for the sample modified by diphenylthiourea (DPTU), which was previously investigated as stabilizer polymer matrix and incorporated therein a laser dye [7]. The level of concentration of the modifying compounds in all cases was the same.

As the table shows, there is a small shift of the position of the fluorescence maximum λ_{fl} in modified samples relative to the check sample. The fluorescence intensity (I) considerably increases at the same time. The maximum value of intensity was detected for the sample of the modified by DTT-2 additive.

In the case of modifying samples with additives DTT, the energy of laser action and lasing photostability of the dye in the polymer matrix increased significantly. At the same time output energy E exceeds the value of corresponding indicator for the polymer matrix modified by DPTU additive.

Table 1. Results of the study of spectral-fluorescent properties and laser test of the samples

Modifying additive*	λ_{fl}, nm	I, a.u.	E_m/E	N_m/N
Without additives	549	1460	-	-
DTT-I	551	2281	8,4	7,3
DTT- 2	553	3385	5,0	4,6
DPTU	550	1892	3,4	4,6

As shown in Fig. 4, the decay time of the fluorescence intensity of the dye in the polymer matrix is approximately the same at all concentrations of the modifying additive - DTT-2. Therefore this type additive does not lead to Rhodamine 6G fluorescence quantum yield dropping in modified polymer matrix.

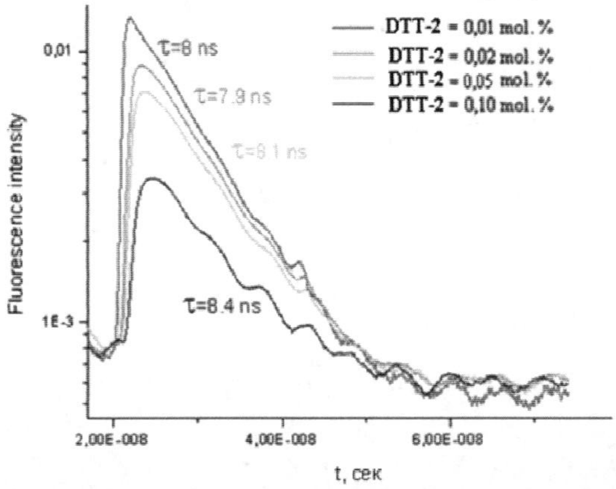

Figure 4. Fluorescence kinetic of the polymer samples doped with Rhodamine 6G and modified by DTT-2.

Journal of Physics: Conference Series **560** (2014) 012017 doi:10.1088/1742-6596/560/1/012017

Conclusion

Therefore, spectral-fluorescent and lasing characteristics of samples of organic glasses, modified by the addition of DTT and doped laser rhodamine 6G dye were studied. There is a tendency to reabsorb the dye fluorescence by studied additives with increasing of their concentration in a polymer matrix. So, it means that necessary to choose the optimal concentration of modifying additives. It is shown that the modification by DTT leads to a considerable increasing in optical density, fluorescence, energy of laser action and the laser photostability of rhodamine 6G dye doped polymer samples, and it may be promising for obtain polymer LAM with improved performances

Acknowledgment

This work partly was funded by the subsidy allocated to Kazan Federal University for the project part of the state assignment in the sphere of scientific activities.

References

[1] Serova V N 2011 *Polymer optical materials* (S.-Petersburg) 384 .
[2] Duarte F J (Ed.) 2009 *Tunable Laser Applications*, 2nd Ed. (CRC) 445
[3] Sunita Singh et al 2003 *Journal of Luminescence* **101** 285
[4] Rânby B 1975 Photodegradation, Photooxidation and Photostabilization of Polymers J.F. Rabek (London)
[5] Krichevskii G E 1968 Fotokhimicheskie prevrashcheniya krasitelei i svetostabilizatsiya okrashennykh materialov 248
[6] Idrisov R et al 2014 *Herald of Kazan Technological University* **17(7)** 148
[7] Serova V N et al 2001 *Russian Polymer New* **6 (3)** 1

Journal of Physics: Conference Series **560** (2014) 012018 doi:10.1088/1742-6596/560/1/012018

Measurements of thermal properties of crystals CaF_2: Ce^{3+} activated Yb^{3+} and Lu^{3+} ions.

N F Rakhimov[1], A S Nizamutdinov[1], V V Semashko[1], P A Popov[2], M A Marisov[1]

[1]Kazan Federal University, 420008, Kremlevskaja str., 18, Kazan, Russian Federation
[2]Petrovskii Bryansk State University, 14, Bezhitskaya str, 241036 Bryansk, Russian federation

E-mail: niyaz-06@mail.ru

Abstract. Aim of this work was to investigate parameters of thermal expansion, thermal conductivity, and the temperature coefficient of the refractive index. These data are very important in the development of optical systems of lasers. Measurements of thermal conductivity were carried out using absolute stationary method of longitudinal heat flux. Parameters of thermal expansion and temperature coefficient of the refractive index were obtained by Jamin and Michelson interferometers.

1. Introduction

Solid solutions with fluorite structure are key components of many nano, poly and nanocrystalline materials, including active and passive photonic materials (solid lasers, scintillators, IR and UV block optics, holography, photolithography and anti-glare coating) and solid electrolytes [1 - 3].

One of the main advantages of fluorite crystals is large spectral large spectral range of transparency which extends to vacuum UV. Good thermal conductivity and low maximum phonon energy provide low probability of nonradiative transitions. The drawback is that charge compensation is needed then doping by trivalent rare earth ions. This could lead to higher probability of defects formation and multiple impurity sites character of crystallization. From another hand this drawback provides an opportunity to achieve larger gain spectrum due to imhomogenious broadening or varying of optical properties [4, 5]. Also it is known that all the rare earth ions exhibit trend to cluster in the fluorite structure. The presence of a large number of clusters of rare earth ions in the doped CaF_2 could be advantageous [6, 7] and is being used to develop an effective system of high power amplifiers and femtosecond lasers [8, 9]. Generally a heterovalent activation nature and clusterization of dopants worsens thermal physical properties of the materials.

Parameters of thermal expansion, thermal conductivity, and the temperature coefficient of the refractive index are the fundamental characteristics of any material. Thermal conductivity determines the temperature distribution over the cross section and over the length of the active element and, consequently, induced thermal lens shape defining the divergence of the laser beam. Particularly it is important to know the thermal conductivity values for evaluating the possibility of applying the crystalline material in a continuous solid state lasers, and pulsed lasers operating at high power levels, due to the problem of heat removal from the active element. In this paper, we present an experimental study of thermal conductivity and temperature coefficient of the refractive index for crystals CaF_2 and solid solutions CaF_2-LuF_3 doped by Yb^{3+} ions.

Journal of Physics: Conference Series **560** (2014) 012018 doi:10.1088/1742-6596/560/1/012018

2. Experimental equipment and measurement technique

Measurements of thermal conductivity were carried out in Bryansk State University named after academician I. G. Petrovsky. As the basis for the design of experimental setup the state special standard unit of thermal conductivity was taken [10]. The scheme of the measuring chamber of the apparatus used in the present work is shown in Figure 1.

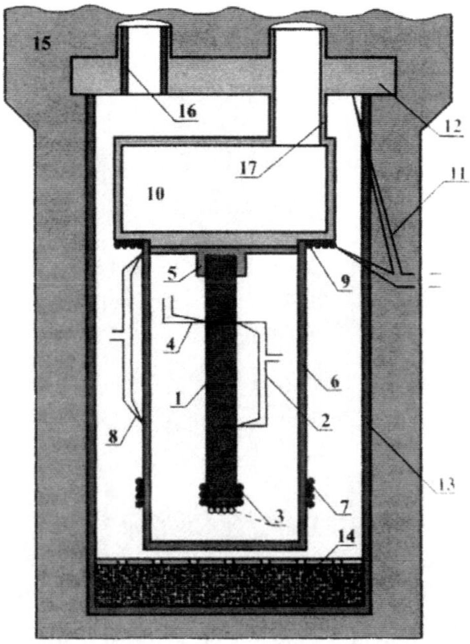

Figure 1. Scheme of the measuring chamber: 1 – investigated sample; 2, 4, 8, 11 - thermocouple; 3,7,9 - heaters; 5 - bushing for mounting the sample; 6 - radiation shield; 10 – heat output; 12 - flange; 13 - evacuation tube; 14 - coal getter; 15 - liquid nitrogen; 16 - radiator for wiring; 17 - thermal resistance

Absolute stationary method of longitudinal heat flux was used in the measurements of thermal conductivity. The samples were prepared in the form of cylinders and were connected to heater and heat output reservoir at the opposite sides. Liquid nitrogen was used as a refrigerant. The heat flux was measured along the sample by means of thermocouple attached to two points of the sample. Measurement procedure is described in detail in [11; 12]. Error in determining the value of thermal conductivity does not exceed ± 5%.

Also, we have determined the temperature coefficient of the refractive index and the coefficient of linear thermal expansion. The scheme of experimental setup utilized is shown in Figure 2.

In order to separate peculiarities arising from thermal expansion of the samples and refractive index change we have utilized the setup consisting of both Michelson and Jamin interferometers.

When illuminating Jamin interferometer with monochromatic light (1 at fig. 2) there are lines of equal inclination at the place of CCD camera 5 appearing due to interference of light beams of branches A and B (see fig. 2). If the wavefront of the light beam is almost flat, the width of the interference fringes exceeds the cross-section diameter of the light beam and the interference pattern illuminates the field uniformly. To observe the interference pattern, the sample is placed in the working beam A.

Journal of Physics: Conference Series **560** (2014) 012018 doi:10.1088/1742-6596/560/1/012018

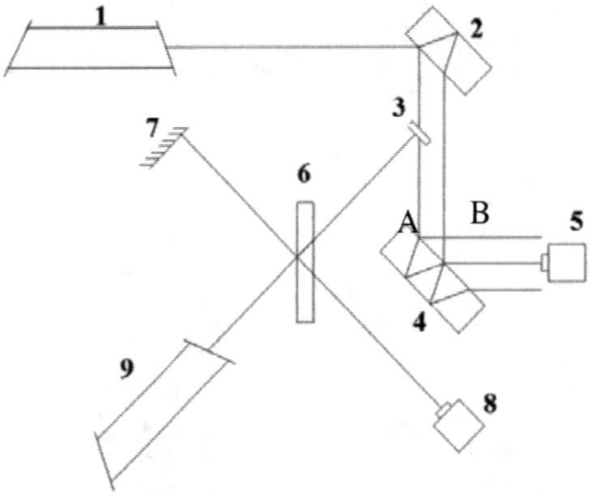

Figure 2. Scheme of interference measurement setup for measuring temperature coefficient of the refractive index and the coefficient of linear thermal expansion: 1 – DPSS laser at 532 nm; 2, 4 – thin (~ 2 cm) plane parallel plates, 6 - plane - parallel plate; 3 - the sample; 5,8 - CCD cameras; 7 - mirror; 9 – He-Ne laser with a wavelength of 632,8 nm, A and B – the working beam and reference beam of Jamin interferometer correspondingly.

Optical path of the laser beam transmitted through the crystal increases when heating the crystal due to its expansion and refractive index changes. When the sample is heated from temperature T_1 to T_2, the optical path length change could be written as follows:

$$\Delta l = n_2(h_1 + dh) - n_1 h_1 - dh \tag{1}$$

where n_1 and h_1 are refractive index and thickness of the sample correspondingly at T_1, n_2 is the refractive index at T_2, *dh* is the value of expansion of the crystal due to temperature change. Because of the change in length of optical path the phase difference between the working and reference beam is varying. This leads to a shift of interference fringes. Number of fringes passed through the field of view of CCD camera due to shift in Jamin interferometer can be determined as:

$$X = \frac{(n_2(h_1 + dh) - n_1 h_1 - dh)}{\lambda} \tag{2}$$

here *X* - number of shifted bands, λ - the wavelength of the light. X could be estimated from counting the number of displaced fringes in the interference pattern, which was filmed by the CCD camera. The refraction index n_1 was determined using a microscope. Contribution of thermal expansion *dh* was determined using a Michelson interferometer. Here we utilized interference of beams reflected from mirror 7 and polished side of the sample 3 (see fig. 2). The sample 3 was mounted at the heater in such a way that one of its polished sides was glued to the heater and other stayed free so that expansion of the sample was equal to the shortening of one of the shoulders of Michelson interferometer.

Heat expansion *dh* of the sample to the value $dh = \lambda_0/4$ (where λ_0 is the wavelength of the He-Ne laser 9) is accompanied by shift of interference pattern and mutual change of places of maximum and minimum intensity. Determining the number of such displacements *N* we can determine the expansion of the sample

$$dh = N\frac{\lambda_0}{4} \tag{3}$$

Unknown member n_2 in formula (2) could be derived from:

$$n_2 = \frac{(X\lambda + n_1 h_1 + dh)}{h_1 + dh} \tag{4}$$

The temperature coefficient of the refractive index and coefficient of linear thermal expansion are determined by the following formulas:

$$n_T = \frac{(n_2 - n_1)}{dT} \quad (5);$$

$$\alpha_T = \frac{dh}{h_1 dT} \quad (6)$$

Temperature of the sample was measured by means of heat-variable resistor.

3. Results

Figure 3 shows the results of measurement of thermal conductivity of the samples. It is seen from the graph that the samples are in the behavior of the thermal conductivity characteristic to glasses, i.e. increase in thermal conductivity with increasing temperature.

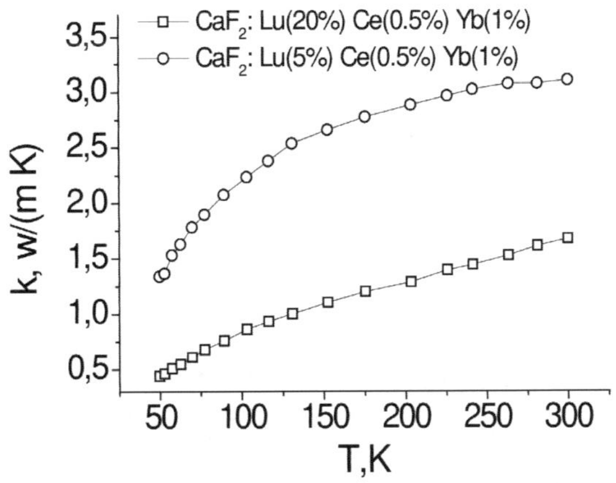

Figure 3. Thermal conductivity **k** of samples of CaF_2-LuF_3 crystals doped with Ce^{3+} and Yb^{3+} ions.

Low thermal conductivity and the weakness of its temperature dependence indicate the presence of inhomogeneities of the crystal structure and other centers of phonon scattering [13, 14]. It also shows that with increasing concentration of LuF_3 thermal conductivity reduces, which is a negative characteristic for the laser material.

Table 1. Temperature coefficient of the refractive index and the coefficient of linear thermal expansion measured for CaF_2-LuF_3 crystals doped with Ce^{3+} and Yb^{3+} ions and CaF_2 table data

	n_T	α_T
CaF_2 (from ref)	-1,0±0,1*10^{-5}	1,8±0,3*10^{-5}
CaF_2: Ce (1 %)	-1,4±0,1*10^{-5}	1,08±0,02*10^{-5}
CaF_2: Ce (0,2 %)+Yb (0,05 %)	-1,5±0,1*10^{-5}	1,24±0,02*10^{-5}
CaF_2-LuF_3 (20%): Ce (1 %)	-2,1±0,4*10^{-5}	1,43±0,07*10^{-5}

In the Table 1 the results of measuring the temperature coefficient of the refractive index and the coefficient of linear thermal expansion are shown. Obtained data show that these parameters experience significant changes when host lattice is doped with RE ions. Although doping level was not high in our samples the temperature coefficient of the refractive index decreases in 1,5 times for Ce^{3+} and Yb^{3+} doped crystals compared to CaF_2 crystal host. It is known that doping CaF_2 crystal by RE ions at levels higher than 0,1 at. % provides significant probability of impurity clusters formation.

4. Conclusion

In present work we have investigated thermal conductivity in the range 50 – 300 K and have estimated temperature coefficient of the refractive index and the coefficient of linear thermal expansion of CaF_2 crystals doped by Ce^{3+} and Yb^{3+} ions and CaF_2-LuF_3 mixed crystal. Thermal conductivity of the investigated materials appears to be lower than that of pure CaF_2 and shows behavior characteristic to glasses, which is the evidence of presence of inhomogeneities of the crystal structure and other centers of phonon scattering due to impurities. The temperature coefficient of the refractive index is also 1,5 times lower for Ce^{3+} and Yb^{3+} doped crystals compared to CaF_2 crystal host. These data are very important in the development of optical systems of lasers.

Acknowledgements

This work was funded by the subsidy allocated to Kazan Federal University for the project part of the state assignment in the sphere of scientific activities.

References

[1] Fedorov P P, Osiko V V, Kuznetsov S V and Garibin EA Fluoride Laser Nanoceramics 2012 *J. Phys. Conf. Ser.* **345** paper 012 017

[2] Fedorov P P, Luginina A A, Kuznetsov S V and Osiko V V Nanofluorides 2011 *J. Fluorine Chem.* **132(12)** 1012–1039

[3] Basiev T T, Doroshenko M E, Fedorov P P et al. Efficient Laser Based on CaF2–SrF2–YbF3 Nanoceramics 2008 *Opt. Lett.* **33(5)** 521–523

[4] Yeqing Wu, Liangbi Su, Qingguo Wang, Hongjun Li, Yaoyu Zhan, Dapeng Jang, Xiaobo Qian, Chuanyong Wang, Lihe Zheng, Hongbing Chen, Jun Xu, Witold Ryba-Romanowski, Piotr Solarz and Radoslaw Lisiecki Spectroscopic properties of Yb-doped CaF_2–YF_3 solid-solution laser crystal 2013 *Laser Phys.* **23** 4

[5] Korner J, Jambunathan V, Hein J, Seifert R, Loeser M, Siebold M, Schramm U, Sikocinski P, Lucianetti A, Mocek T, Kaluza M C Spectroscopic characterization of Yb^{3+} doped laser materials at cryogenic temperatures 2014 *Appl. Phys. B* **116** 75–81

[6] Petit V, Camy P, Doualan J L, Portier X and Moncorgé R Spectroscopy of Yb3+:CaF2: From isolated centers to clusters 2008 *Phys. Rev. B* **78** 085131

[7] Chernyshev V A, Nikiforov A E, Volodin V P and Slepukhin G S Electronic structure of Yb3+ impurity centers in fluorites 2010 *Phys. Solid State* **52** 1874–1879

[8] Siebold M, Bock S, Schramm U, Xu B, Doualan J L, Camy P and Moncorgé R Yb:CaF2-a new old laser crystal 2009 *Appl.Phys.B* **97** 327–338

[9] Ricaud S, Druon F, Papadopoulos D N, Pellegrina A, Balembois F, Georges P, Courjaud A, Camy P, Doualan J L and Moncorgé R High-power diode-pumped cryogenically cooled Yb:CaF2 laser with extremely low quantum defect 2011 *Opt. Lett.* **36** 1602–1604

[10] State special standard unit of thermal conductivity of solids in the temperature range 4.2-90 K. 1984 *Report of the USSR State Committee on Standards* 26

[11] Popov P A Thermal conductivity of laser crystals with garnet structure in the temperature range 6 - 300 K: dis. ... 1993 *Cand. nat. - mat. Sciences* 214

[12] Sirota N N, Popov P A, Ivanov I A The thermal conductivity of monocrystalline gallium garnets doped with rare-earth elements and chromium in the range 6 – 300 K 1992 *Cryst. Res. Technol* **27(4)** 535-543

Distribution coefficient of Pr^{3+} ions in crystals of solid solutions LiF-LuF_3-YF_3-PrF_3

A A Shavelev, A S Nizamutdinov, V V Semashko, S L Korableva, V G Gorieva
Kazan Federal University, 420008, Kremlevskaja str., 18, Kazan, Russian Federation

E-mail: alexey.shavelev@gmail.com

Abstract. Aim of this work was to investigate the possibility of increasing the distribution coefficient of Pr^{3+} ions in fluoride crystals with scheelite structure. As a result of investigations by XRF and optical absorption spectroscopy the absorption cross section for $^3H_0 \rightarrow {}^3P_0$ transitions Pr^{3+} ions was clarified and segregation coefficient of Pr^{3+} ions in $LiY_{0,3}Lu_{0,7}F_4$ mixed crystals was estimated.

1. Introduction

In recent years a considerable interest in the development of efficient laser sources emitting in the visible spectral range with laser diode pumping occurred. Prospects for such devices lay in development of color displays, projectors, as well as biomedical and quantum-optical applications and industry [1, 2]. Development of laser diodes technology provides increase of number of works on Pr^{3+} based active media [3,4,5]. One of the most promising materials for such active media is $LiYF_4$ crystal, activated with Pr^{3+} ions. For example, in the work [6] two InGaN laser diodes emitting at λ_p=443,9 nm were used to excite the $^3H_4$-$^3P_2$ transition of Pr^{3+} in $LiYF_4$. The differential efficiency of 54% has been demonstrated in the green region of the spectrum at 522,6 nm. The laser oscillation at wavelengths 523, 546, 604, 607, 640, 698, and 720 nm was achieved in Pr:$LiYF_4$ as described in the work [7] and the output power of 2,9 W and slope efficiency of 72 % have been shown for 532 nm lasing with 5 W frequency doubled optically pumped semiconductor laser pump at 479 nm.

In present work we investigate the possibility of increasing the distribution coefficient of Pr^{3+} ions in fluoride crystals with scheelite structure, which is an actual problem [8]. Earlier in Kazan Federal University it was found that the transition from $LiYF_4$ and $LiLuF_4$ crystals to their solid solutions $LiY_{1-x}Lu_xF_4$ increases sharply the distribution coefficient of rare-earth ions (REI), which leads to increased concentrations of rare-earth ions and improvement the optical quality of the material [8].

2. Object of research, scheme and experimental techniques

Objects of the study were solid solutions LiF-LuF_3-YF_3-PrF_3. Three series of crystals with chemical compositions $LiLuF_4$:Pr^{3+} (0,5 at. %), $LiLu_{0,7}Y_{0,3}F_4$:Pr^{3+} (0,5 at. %) and $LiLu_{0,7}Y_{0,3}F_4$:Pr^{3+} (1 at. %) were synthesized. The crystals were grown in carbon crucibles by the Bridgman-Stockbarger method. Because of irregular distribution of impurities in the samples along the crystal boule different groups of samples were prepared for X-ray fluorescence and optical spectroscopy studies: 3 plane-parallel plate samples of $LiY_{1-x}Lu_xF_4$: Pr^{3+} (0,5%) cut from different places of the crystal boule with respect to the start of crystallization and the same series were prepared for the sample $LiY_{1-x}Lu_xF_4$: Pr^{3+} (1%). Also plane-parallel plate sample was prepared from the middle of $LiLuF_4$ boule. For all cases the optical c-axis was oriented in the plane of polished windows. The experimental scheme for the absorption spectra detection is shown in Figure 1.

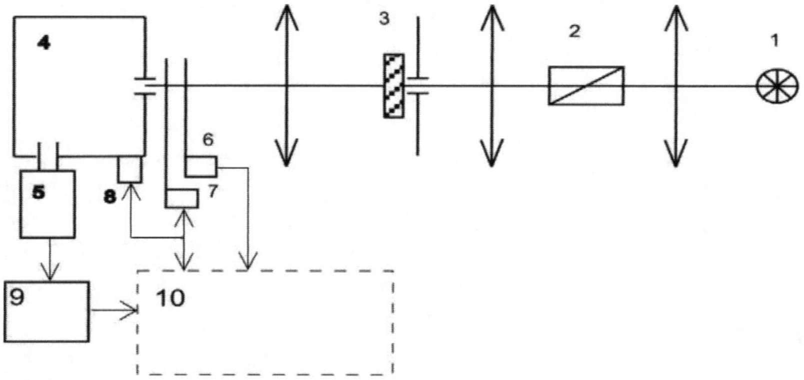

Figure 1. Experimental setup: 1 - source of the continuous spectrum; 2 - Glan prism; 3 - diaphragm and the sample; 4 - monochromator MDR-23; 5 - photomultiplier tube; 6 - luminous flux modulator (chopper); 7 – optomechanical shutter; 8 - stepper motor for scanning the spectrum range; 9 - selective amplifier; 10 - ADC and computer.

3. Research Methods

Information on the actual concentration of the impurity ions in the synthesized single crystals can be obtained by X-ray fluorescence analysis (XRF) and optical absorption analysis.

XRF method is based on detection of transitions between the deep-lying energy levels of the atom, when electrons are always localized at the inner shells of the atom and give the information on real concentration of the given chemical element. Absorption spectroscopy method is characterized by detection of transitions between the outer energy levels of the atom and it is used to evaluate the perspective of new materials.

The investigated fluoride crystals with scheelite structured crystal lattice are characterized by low isomorphic capacity in respect to rare-earth ions of the first half of the row due to significant mismatch of ionic radii of cations substituting each other. This means that segregation coefficient of these ions will be less than unity, e. i. not all the PrF_3 component we put to the charge appears in the crystal lattice of synthesized material. The resulting solid concentration $C_S(x)$ was measured by means of XRF spectroscopy. Knowing the real concentration of the impurity ions and absorption coefficient we can obtain the cross section of the transition value.

Thus the absorption cross section could be estimated according to the following expression:

$$\sigma(\lambda) = \frac{\kappa(\lambda)\cdot\mu}{C_S\cdot\rho\cdot Na} \tag{2}$$

Where k - the absorption coefficient, μ - the molar mass of the crystal, σ - the absorption cross section, ρ – density, Na - Avogadro's number, C_s - Pr^{3+} ions solid concentration (at. %).

4. Experiment Results

The X-ray diffraction was studied from grown crystalline material in order to investigate the phase composition of the mixed crystals.

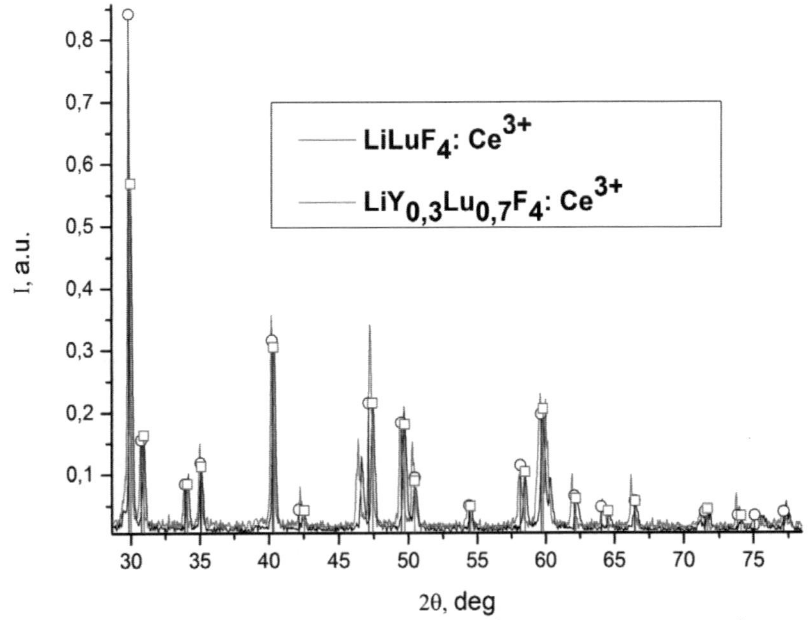

Figure 2. X-ray diffraction results of the samples: $LiLuF_4$:Ce^{3+} and $LiY_{0,3}Lu_{0,7}F_4$: Ce^{3+}

The diffraction pattern obtained for Ce^{3+}: $LiY_{0,3}Lu_{0,7}F_4$ (see Fig. 2) corresponds to scheelite structure. For transition from $LiLuF_4$ to $LiY_{0,3}Lu_{0,7}F_4$ data the shift is observed towards bigger angles due to larger lattice constant, as it was expected.

The measured concentration of Pr^{3+} ions C_s dependent on place at the crystal boule is presented in Table 1. The data for $LiLuF_4$:Pr^{3+} sample corresponds to the hat of the boule. It is seen that real concentration of Pr^{3+} ions is slightly higher for mixed crystal than that for $LiLuF_4$ at the same level of concentration in the initial charge.

Table 1. Concentration of the samples: a) $LiLuF_4$:Pr^{3+}(0,5%); b) $LiY_{0,3}Lu_{0,7}F_4$:Pr^{3+}(0,5%); c) $LiY_{0,3}Lu_{0,7}F_4$:Pr^{3+}(1%).

Sample	C_s, at. % (from XRF data)	L, mm (point from crystallization start)
a) $LiLuF_4$:Pr^{3+}	0,25	23
b) $LiY_{0,3}Lu_{0,7}F_4$:Pr^{3+} (HAT)	0,31	40
b) $LiY_{0,3}Lu_{0,7}F_4$:Pr^{3+} (MID)	0,2	23
b) $LiY_{0,3}Lu_{0,7}F_4$:Pr^{3+} (NOSE)	0,16	6
c) $LiY_{0,3}Lu_{0,7}F_4$:Pr^{3+} (HAT)	0,38	40
c) $LiY_{0,3}Lu_{0,7}F_4$:Pr^{3+} (MID)	0,3	23
c) $LiY_{0,3}Lu_{0,7}F_4$:Pr^{3+} (NOSE)	0,24	6

Journal of Physics: Conference Series **560** (2014) 012019 doi:10.1088/1742-6596/560/1/012019

In order to determine segregation coefficient of Pr^{3+} ions in mixed crystal $LiY_{0,3}Lu_{0,7}F_4$ the dependence of ln[1-C_s/C_0] on length of the crystal was plotted (see Fig. 3). Slope of the dependence gives the value of segregation coefficient.

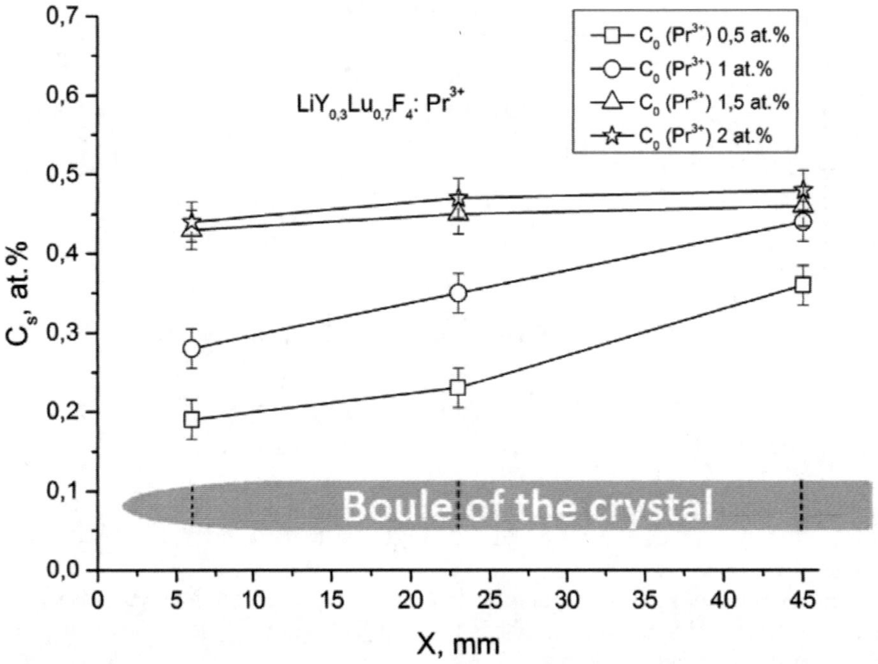

Figure 3. Longitudinal distribution of the resulting solid concentration $C_S(x)$ of praseodymium ions in the samples of crystals of solid solutions $LiY_{0,3}Lu_{0,7}F_4$:Pr^{3+}.

The Figure 3 shows that, as crystals are grown in a closed type crucible, the concentration of rare earth ions increases in the melted part during crystal growth. For the middle part of the boule that is usually associated with higher optical quality of the synthesized crystalline material real Pr^{3+} ions concentration was $0,4*10^{20}$ cm^{-3}. It is seen that segregation coefficient decreases with increase of concentration of Pr^{3+} in the initial charge C_0. This is the evidence of finite isomorphic capacity of $LiY_{0,3}Lu_{0,7}F_4$ host lattice. For the cases of C_0 levels of 1,5 at. % and 2 at. % the dependence on point at the crystal is almost vanished and this can lead to formation of significant amount of defects inside the crystal.

Registered absorption spectra in the middle of Pr:$LiY_{0,3}Lu_{0,7}F_4$ crystals boules in π-polarization are presented in Figure 2. The observed line at 479,2 nm corresponds to the $^3H_4 \rightarrow {}^3P_0$ transition of praseodymium ions. The absorption coefficient at the maximum has value of 3,5 cm^{-1} for crystal with 1 at. % of Pr^{3+} in the initial charge and this line is prospective for pumping the Pr^{3+} ions by means of laser diode light. Knowing the real value of Pr^{3+} concentration in the crystal we have obtained the value of absorption cross section at the wavelength of 479,2 nm which has value of $12 \pm 3 * 10^{-20}$ cm^2.

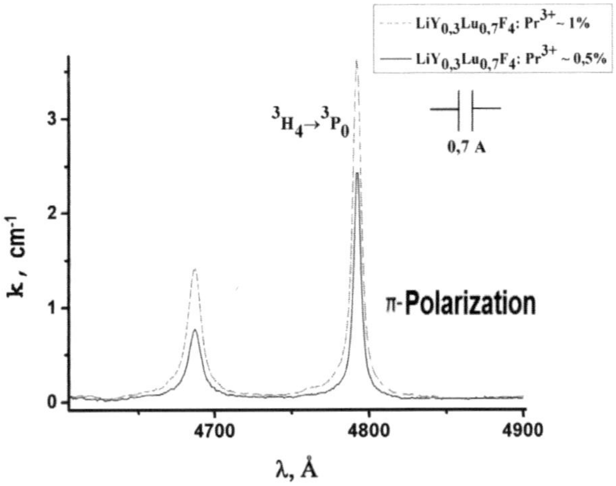

Figure 2. The absorption spectra of crystals and solid solutions of $LiLu_{0,7}Y_{0,3}F_4$: Pr^{3+} (0,5%) and $LiLu_{0,7}Y_{0,3}F4$: Pr^{3+} (1%).

5. Conclusion

Here we have investigated the segregation coefficient of solid solutions of scheelite structure LiF-LuF_3-YF_3-PrF_3 in respect to Pr^{3+} ions. It has been shown that segregation coefficient decreases with increase of concentration of Pr^{3+} in the initial charge. Low isomorphic capacity of the investigated host lattice in respect to Pr^{3+} ions leads to saturation of segregation coefficient at the 1,5 at. % and 2 at. % of Pr^{3+} in the initial charge. Thus in order to obtain crystals of good optical quality by growing with Bridgeman technique it is proposed to use 0,5 or 1 at. % of Pr^{3+}. The real concentration of Pr^{3+} for such charges appears to be 0,3 at. % ($0.4*10^{20}$ cm^{-3}) which is slightly higher than that for Pr:$LiLuF_4$ crystal. Also we were able to clarify the absorption cross section of $^3H_0 \rightarrow {}^3P_0$ transitions of Pr^{3+} ions in investigated crystals which has the value of $12 \pm 3 * 10^{-20}$ cm^2. Thus, the new crystals are promising as active media for diode-pumped laser.

Acknowledgements
This work was funded by the subsidy allocated to Kazan Federal University for the project part of the state assignment in the sphere of scientific activities.

References

[1] Conchello J-A and Lichtman J W 2005 *Nat. Methods* **2** 920
[2] Engler S, Ramsayer R and Poprawe R 2011 *Phys. Procedia* **12** 339
[3] Metz P W, Müller S, Reichert F, Marzahl D-T, Moglia F, Kränkel C, Huber G 2013 *Optics Express* **21 (25)** 31274-31281
[4] Fujimoto Y, Nakanishi J, Yamada T, Ishii O, Yamazaki M *Progress in Quantum Electronics* **37 (4)** 185-214
[5] Fibrich M, Jelínková H, Šulc J, Nejezchleb K, Škoda V 2011 *Laser Physics Letters* **8 (8)** 559-568
[6] Metz P W, Reichert F, Moglia F, Müller S, Marzahl D-T, Kränkel C, Huber G. 2014 *Optics Letters* **39 (11)** 3193-3196
[7] Richter A, // 2008 *Cuvillier Verlag*
[8] Nizamutdinov A S, Semashko V V, Naumov A K, Efimov V N, Korableva S L, Marisov M A, 2010 *JETP Letters* **91** 21-23

Institute of Physics
Dirac House, Temple Back
Bristol BS1 6BE UK

CURRAN ASSOCIATES INC
proceedings.com